T5-CPY-539

OPTIMAL VLSI ARCHITECTURAL SYNTHESIS

Area, Performance and Testability

THE KLUWER INTERNATIONAL SERIES IN ENGINEERING AND COMPUTER SCIENCE

VLSI, COMPUTER ARCHITECTURE AND DIGITAL SIGNAL PROCESSING

Consulting Editor
Jonathan Allen

Latest Titles

Low-Noise Wide-Band Amplifiers in Bipolar and CMOS Technologies, Z.Y.Chang, W.M.C.Sansen,
ISBN: 0-7923-9096-2

Iterative Identification and Restoration of Images, R. L.Lagendijk, J. Biemond
ISBN: 0-7923-9097-0

VLSI Design of Neural Networks, U. Ramacher, U. Ruckert
ISBN: 0-7923-9127-6

Synchronization Design for Digital Systems, T. H. Meng
ISBN: 0-7923-9128-4

Hardware Annealing in Analog VLSI Neurocomputing, B. W. Lee, B. J. Sheu
ISBN: 0-7923-9132-2

Neural Networks and Speech Processing, D. P. Morgan, C.L. Scofield
ISBN: 0-7923-9144-6

Silicon-on-Insulator Technology: Materials to VLSI, J.P. Colinge
ISBN: 0-7923-9150-0

Microwave Semiconductor Devices, S. Yngvesson
ISBN: 0-7923-9156-X

A Survey of High-Level Synthesis Systems, R. A. Walker, R. Camposano
ISBN: 0-7923-9158-6

Symbolic Analysis for Automated Design of Analog Integrated Circuits, G. Gielen, W. Sansen,
ISBN: 0-7923-9161-6

High-Level VLSI Synthesis, R. Camposano, W. Wolf,
ISBN: 0-7923-9159-4

Integrating Functional and Temporal Domains in Logic Design: The False Path Problem and its Implications, P. C. McGeer, R. K. Brayton,
ISBN: 0-7923-9163-2

Neural Models and Algorithms for Digital Testing, S. T. Chakradhar, V. D. Agrawal, M. L. Bushnell,
ISBN: 0-7923-9165-9

Monte Carlo Device Simulation: Full Band and Beyond, Karl Hess, editor
ISBN: 0-7923-9172-1

The Design of Communicating Systems: A System Engineering Approach, C. J. Koomen
ISBN: 0-7923-9203-5

Parallel Algorithms and Architectures for DSP Applications, M.A. Bayoumi, editor
ISBN: 0-7923-9209-4

Digital Speech Processing: Speech Coding, Synthesis and Recognition, A. Nejat Ince, editor
ISBN: 0-7923-9220-5

Assessing Fault Model and Test Quality, Kenneth M. Butler, M. Ray Mercer
ISBN: 0-7923-9222-1

OPTIMAL VLSI ARCHITECTURAL SYNTHESIS

Area, Performance and Testability

Catherine H. Gebotys
and
Mohamed I. Elmasry

University of Waterloo

Kluwer Academic Publishers
Boston/Dordrecht/London

Distributors for North America:
Kluwer Academic Publishers
101 Philip Drive
Assinippi Park
Norwell, Massachusetts 02061 USA

Distributors for all other countries:
Kluwer Academic Publishers Group
Distribution Centre
Post Office Box 322
3300 AH Dordrecht, THE NETHERLANDS

Library of Congress Cataloging-in-Publication Data

Gebotys, Catherine H.
Optimal VLSI architectural synthesis : area, performance, and testability / Catherine H. Gebotys and Mohamed I. Elmasry.
p. cm. -- (The Kluwer international series in engineering and computer science. VLSI, computer architecture, and digital signal processing)
Includes bibliographical references and index.
ISBN 0-7923-9223-X (alk. paper)
1. Computer architecture. 2. Integrated circuits--Very large scale integration. I. Elmasry, Mohamed I., 1943- . II. Title. III. Series.
QA76.9.A73G42 1992
004.2'2--dc20

91-31898
CIP

Printed on acid-free paper.

Printed in the United States of America

To

Robert Joseph and Kathleen Vanessa Gebotys

and

Elizabeth, Carmen, Samir, Nadia and Hassan Elmasry

Table of Contents

PREFACE

Although research in architectural synthesis has been conducted for over ten years it has had very little impact on industry. This in our view is due to the inability of current architectural synthesizers to provide area-delay competitive (or "optimal") architectures, that will support interfaces to analog, asynchronous, and other complex processes. They also fail to incorporate testability. The OASIC (optimal architectural synthesis with interface constraints) architectural synthesizer and the CATREE (computer aided trees) synthesizer demonstrate how these problems can be solved.

Traditionally architectural synthesis is viewed as NP hard and therefore most research has involved heuristics. OASIC demonstrates by using an IP approach (using polyhedral analysis), that most input algorithms can be synthesized very fast into globally optimal architectures. Since a mathematical model is used, complex interface constraints can easily be incorporated and solved.

Research in test incorporation has in general been separate from synthesis research. This is due to the fact that traditional test research has been at the gate or lower level of design representation. Nevertheless as technologies scale down, and complexity of design scales up, the push for reducing testing times is increased. On way to deal with this is to incorporate test strategies early in the design process. The second half of this text examines an approach for integrating architectural synthesis with test incorporation. Research showed that test must be considered during synthesis to provide good architectural solutions which minimize area delay cost functions.

Though originally developed separately, OASIC and CATREE can be integrated so that OASIC simultaneously schedules and allocates the architecture and CATREE performs binding (and reallocating) of the architecture for testability.

Part I introduces the motivation and current open problems with high level CAD. Part II provides the necessary background material on architectural synthesis and integer programming. This part includes a definition of problems in both areas and a brief review of previous approaches to solving these problems. Part III outlines the OASIC methodology, models, the solution techniques used, and some synthesized results. Part IV outlines the CATREE methodology, the algorithms and data structures used and some synthesized results. Part V provides a brief discussion and concluding remarks concerning how we will interface with CAD tools of the future.

The book can be used at the senior undergraduate and graduate levels in courses dealing with computer architectures, computer organization, VLSI design, computer-aided design, VLSI digital signal processing, testing, or integer programming. It will be also of value to resesarchers dealing with these topics.

C.H. Gebotys
M.I. Elmasry
Waterloo, Ontario, Canada

PART I : INTRODUCTION

1.

GLOBAL VLSI DESIGN CYCLE

The global VLSI (very large scale integration) systems design cycle is briefly discussed below with respect to relationships between design stages, bottlenecks, and current open issues for design automation (DA). The design cycle involves moving from an abstract design specification to gradually a more detailed single or multichip design that can be tested and fabricated. The VLSI design stages are very interdependent and therefore it is important to outline the purpose of each stage before one can address the problems of high level synthesis. Area, power, speed, timing issues, input and output pin limitations, testability, and many other criteria are important in the design process. Interfaces to other complex processes, design complexity with respect to implementation technologies and testability will also be discussed. In addition, an understanding of the current computer aided-design (CAD) bottlenecks and open issues will further emphasize the importance and impact of high level architectural synthesis (the focus of this text) on the VLSI design cycle.

1.1 VLSI DESIGN CYCLE

The VLSI systems design cycle generally involves many transformations from a high level design specification to a low level of design representation. Stages include, but are not limited to : local and global transformations on the behavioral specification, partitioning of the behavior, architectural synthesis (transforming behavior into an architecture), logic synthesis, functional level simulation, module generation, placement and routing, timing analysis, and final mask layout and verification. The behavioral specification, also called the input algorithm, which is accompanied by a cost function that drives the design synthesis. For example the cost function may involve the minimization of chip area and power dissipation, or the maximization of chip speed and testability.

The time required for each design stage may be quite large depending upon how much automation is provided or the designers expertise. Feedback from one stage to a previous stage is often quite frequent and time consuming due to incorrect early decisions or false assumptions. For example a partitioning decision may lead to a chip which exceeds its area requirements and therefore feedback is required to correct the earlier partition decision. Feedback is often inevitable since each design cycle step is interdependent upon the others. For example a decision made during behavioral scheduling affects all lower stages such as hardware allocation and the final VLSI design layout. Yet it is very difficult to predict the effects that early decisions will have since the behavior is technology independent. It is believed to be impossible to simultaneously consider all stages (down to layout) due to the complexity and enormous amounts of data required. It is well known that early decisions made in the design cycle often have the greatest impact on the final design. Thus the high level stages are currently viewed as being very important and of great interest in the VLSI community.

Early steps of the VLSI design cycle have been defined as algorithm transformations, algorithm partitioning, and architectural synthesis. We will use the terms behavior or algorithm to describe the input into an architectural synthesizer. Decisions to partition the behavior among multiple chips (spatial), different analog and digital domains (technology), or into separate pipestages (time) are explored in these early steps. In industry these decisions are often done without the aid of design automation tools, yet it is this exploration which is considered critical for shortening the design cycle time and of great importance for designing high performance architectures.

Even though research on high level architectural synthesis tools has been conducted for more than ten years, it has not had a significant impact on industry. This can be attributed to the known fact that the acceptance of new technologies occurs much faster in industry (Langeler, 1989) than the acceptance of new DA tools. Currently the most common and mature DA tools in industry perform low level tedious tasks such as module generation, placement, routing, and layout. Figure 1.1 illustrates the maturity of the various DA tools. More recently logic synthesis tools have been introduced into the CAD market for controller design. We believe there are several reasons why architectural synthesis and higher level tools have not found a place in industry. In order to understand why, we will first briefly introduce the subject of architectural synthesis, and then look at issues which have not been adequately addressed by researchers and consequently contribute to preventing the introduction of synthesis in industry. Chapters three and four, in part II, will review the field of architectural synthesis and integer programming respectively. In chapter five through nine, of part III, we will introduce our formal and practical approach to solving these issues optimally for architectural synthesis.

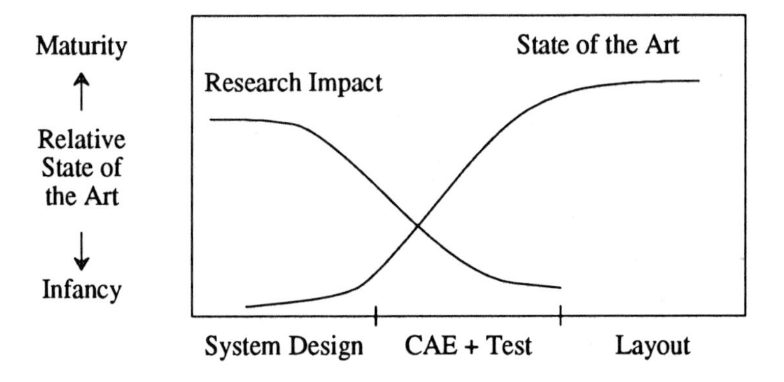

Figure 1.1. The maturity of various CAD tools for VLSI.

Architectural synthesis of digital synchronous chips refers to the transformation from a behavioral (or algorithmic) input description to a hardware architecture which implements the behavior (according to a schedule). The Y chart (Gajski, 1983) shown in figure 1.2, is most commonly used to represent the transformations performed during the design cycle. The three axis of the Y chart, behavioral, structural and geometrical axes, are used to represent different levels of design representation and the mappings required to design a chip. During the design of a system one starts with a behavioral design specification (in theory) and moves successively down the chart to refine the design into greater levels of detail. However we will give a brief look at the different levels of hierarchy by starting at the lowest level and moving up.

Starting at the inner bold dot in figure 1.2 on the behavioral axis, the lower level cells, such as a data storage element is defined. By moving up to the structural axis it becomes more refined as an interconnection of

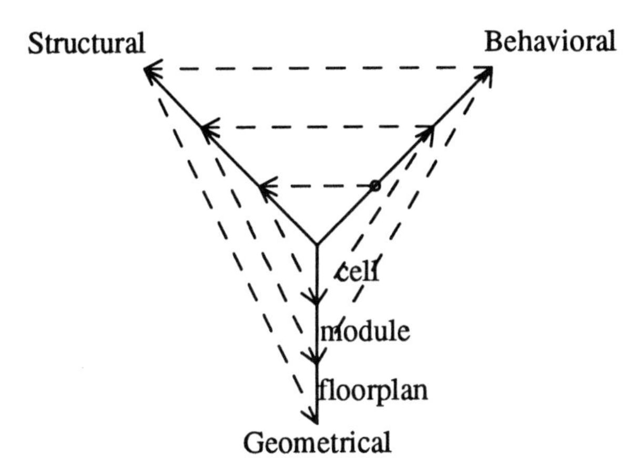

Figure 1.2. The Y chart, illustrating design cycle transitions. The architectural synthesis is represented by the arc, at the top of the chart, from the behavioral axis to the structural axis.

transistors which implements for example a master-slave flip-flop or register. Finally a transition to the geometrical axis defines the actual mask layers required to fabricate the register as part of a chip. When all cells are available in a library one can continue to the module level. Modules are larger design components that are made from a number of cells. An example of a module is a register file, which is composed of a number of registers and control circuitry. The behavioral axis, at the module level, defines the control and functionality of the register file. By moving again to the structural axis the definition of this module in terms of an interconnection of cells is detailed. By moving again to the geometrical axis the module is defined according to the mask definitions of its cells and their interconnection. One more level up the hierarchy defines the behavior of a larger system of modules which could be a chip or system.

The transition from the behavioral axis to the structural axis, at the top of figure 1.2, represents architectural synthesis, the focus of this text. After performing architectural synthesis, a digital designer moves from the structural axis to the layout axis at the module and chip levels during the design cycle. This later procedure is now very well defined in industry. For example a behavior may be described as a z-diagram for digital filters or a programming language (such as 'C' or Pascal). We define a *code operation* as a single specific arithmetic operation in the algorithm. Code operations in the behavioral description may receive data or transfer data in the form of bits, bytes, arrays or strings for computation. Additionally there may be very complex timing constraints on data transfers or communication with external processes. The term *external process* will be used to describe a process or operation that communicates with the behavioral algorithm but is not being synthesized. The output of the high level architectural synthesizer is a hardware architecture and a schedule. An architecture is composed of hardware components, which include registers (memory), busses, and functional units (such as multipliers, arithmetic logic units (ALUs), etc). The architectural synthesis involves many tasks including scheduling and allocation. The schedule defines the mapping of code operations to control states. The allocation tasks determine the number of functional units, registers, and busses. The binding task defines the mapping of code operations to hardware components (functional units and registers), including data transfers to and from busses. Since synthesis is a one to many mapping, often a set of design constraints or a cost function are specified by the user to select among the design solutions or find the optimal one. For example each code operation can be mapped to many different control states and hardware components. The design constraints most often include area and speed (McFarland, 1986) , however other constraints such as power (Petersen, 1986, Haroun, 1989) and test (Gebotys, 1989) , may also be important.

One of the major purposes of architectural design synthesis is to decrease the VLSI design cycle time. In effect, because designs are synthesized faster (than humans can design) there is more time for design exploration and thus yielding 'better' or optimal architectures (with respect to other solutions produced by the synthesizer during design exploration). This tool also provides a good method for handling last minute design specification changes, since new architectures can be quickly synthesized.

Chip level design synthesis is viewed as an important stage in the VLSI design cycle. It follows the design specification stage, where designers define exactly what function their system will perform and how it will be partitioned among custom chips. It also precedes the layout stage. Since the synthesis stage, which determines the architecture of the design is estimated to be 30-40% of the total design effort (Fey, 1986) , automation plays an important part in saving time and manpower.

As VLSI technologies scale up to ULSI (ultra-large-scale-integration) levels the computational demands placed on DA tools increases. This burden affects synthesis tools directly. Behavioral design descriptions to be synthesized will be extremely complex and large. Very few synthesizers have synthesized more than 1000 lines of input code. Only synthesizers targeted for microprocessor designs (Rajan, 1989) have produced architectures for the M68000, using 2426 lines of input code. Most have used far less than 1000 lines of input code to synthesize examples (see the high level synthesis benchmarks at electronic address hlsw@decwrl.dec.com (Borriello, 1988)). It is not clear what limits these architectural synthesis tools exhibit as design sizes increase. Furthermore some subtasks associated with design synthesis, have been classified as NP-hard (Garey, 1979) . This means that there will exist some problems that will require exponential time to solve. In the future, better algorithmic techniques to handle the complexity of the problem will be developed, such as partitioning and execution of synthesis tasks on multiprocessor architectures. The high level synthesis tools must be able to synthesize architectures from partitioned code segments and from input algorithms with a high degree of regularity. These techniques will provide solutions to handling the computational complexities and demands of large systems that need to be synthesized.

1.2 HYBRID SYSTEMS DESIGN

Lacking even more automation is analog and asynchronous design, although recent research has shown much promise for both areas. It is believed that about 30% of ASICS have analog components and by the year 2000 this number is expected to double (Carley, 1989) . Mixed analog/digital systems design may involve tightly coupled (embedded) hardware or loosely coupled (partitionable) mixed hardware components. An example of the former case is the implementation of the artificial neural network algorithm (ANN) where both analog and digital components can be used for different processing aspects. In some cases the advantages of analog or digital implementation may not be clearly identifiable for a particular application and design exploration will be extremely important to identify the optimized combination of digital signal processing (DSP) and analog signal processing (ASP). Loosely coupled mixed designs more commonly occur when a sequential pipeline of processing functions are synthesized. In these types of designs often the partition between analog and digital is well defined. For example a design to drive an RGB (red-green-blue) display may have DAC (digital to analog converter) circuitry on the same chip that provides digital graphics processing. It is believed that high level synthesis of analog and digital circuits is different enough to necessitate the use of separate CAD tools. Nevertheless the research in analog CAD tools greatly lags digital CAD tool research and high level synthesis tools have not been defined in analog design. However, it is important to have a formally defined interface between analog and digital so that the concurrent execution of synthesizers can be performed. We define an *analog interface* to a synchronous digital circuit as a sequential synchronous data input/output at a fixed rate. For example data input to or output from a DAC or a ADC may be a part of the system behavior which is input to an architectural synthesizer. This model of interface will be discussed more in chapter 2.2.

Asynchronous designs are expected to increase due to the limits of global clocking of synchronous circuits including clock skew (Meng, 1989) . Asynchronous circuits have task dependent or data dependent completion times. For example the next task cycle is started once the current task is completed. Asynchronous designs can be represented as bounded or unbounded delay circuits (Meng, 1989) . We define an *asynchronous interface* of a digital synchronous system as inputs synchronized with the controllers (global) clock but are still indeterminate with respect to the control state (or control step) of the system (Hayati, 1989) . We call the interface *bounded* if an earliest and a latest control state is defined. Thus it is known that input data from the external asynchronous process will arrive at a control state greater than or equal to the earliest state and less than or equal to the latest state. Analogously the data could be output to an asynchronous process. In this case, the interface would be used to control the use of the register hold time for transferring output data. There are a number of designs for interfacing asynchronous circuits to digital synchronous systems such as data detectors, spacers, multivalued circuits or other types of synchronizers. The design of these synchronizer components will not be addressed however more information on these can be found in (Balraj, 1986, Brzozowski, 1990, Meng, 1989) .

In totally synchronous digital multichip designs, it may be very difficult to guarantee that the clock signal will arrive at the same time at all parts of the circuits (Brzozowski, 1990) . Thus external data inputs may be delayed by different amounts. Hence it is often necessary to consider asynchronous behavior, even in totally distributed synchronous systems. As geometries scale down, clock skew, slower transition times, and the capacitance and resistance effects will become increasingly dominant (Subrahmanyam, 1988) . Preliminary analysis indicates that for large designs self timed disciplines may be necessary. Additionally in globally synchronous circuits there may also be asynchronous behavior when access to a shared resource is requested. Finally, the systems design may involve loosely coupled VLSI circuits which can be locally synchronous but globally self timed and therefore require asynchronous behavioral interfaces. Asynchronous designs offer many advantages such as increased operating speeds and they can be designed to handle bounded metastable states.

The focus of this text will be on the automated architectural design of digital synchronous circuits with interface to both analog and asynchronous circuits.

1.3 IMPACT OF TECHNOLOGIES

There are many technological factors which also may drive the performance of VLSI designs. For example not only will constraints for data transfer between multichips on a printed circuit board (PCB) board be important for architectural synthesis but also data transfers on a silicon substrate (multichip modules) (Weber, 1989) , or on a wafer (wafer scale integration) will be important. An example of these data transfer constraints are die to die communication delays. Each new technology brings a new set of constraints which must be incorporated into automated architectural synthesis tools.

In addition to the medium of data transfer between chips, the implementation technology of the chip itself is also important. For example the area and delay characteristics of the module library will change when new technologies are introduced. Thus architectural synthesizers must have adaptable cost functions which will take these changes into consideration in finding optimal architectural solutions. These factors will also have a great impact on the types of architectures which may be suited for a technology. For example if interconnect is very expensive (ie. maybe the technology only allows two levels of metal) then busses must be minimized and used to yield a more efficient solution than the use of a random topology (ie. local interconnections).

Another technological impact which affects architectural synthesizers is at the application end. For example artificial neural network (ANN) algorithms are being used in many systems applications such as pattern recognition (Treleavan, 1989) . The ability to embed these algorithms in a systems design has become very important. Systolic (Kung, 1988) and multiprocessor network implementations have already been investigated. Their use as input to architectural synthesizers however has not been explored. For example the VLSI implementation of the ANN may be one part of a larger design of analog signal processing (ASP) or digital signal processing (DSP). It is possible that the execution of the ANN does not conflict in time with other DSP postfiltering and thus the sharing of hardware may be possible.

Architectural synthesizers must be able to handle a wide range of algorithms. These types of algorithms have not been input to existing architectural synthesizers and are characterized by an extremely large number of data transfers between code operations. This factor leads to extremely large interconnect requirements. Since most synthesizers deal with interconnect during the final stages of the design it is possible that they would output unsuitable architectures (where the interconnect complexity is too high). The architectural synthesizer we will present can optimize interconnect at an early stage.

1.4 TEST CONSIDERATIONS

Test is required to verify that the fabricated VLSI chip or multichip system works fault-free or operates satisfactorily (McCluskey, 1990) . In other words a test set or set of test vectors is used to detect faults present in the chip. These faults may be due to the fabrication process or layout errors. Test is not the same as design verification. Design verification, refers to proving that the synthesized design solution (not fabricated chip) is correct with respect to the behavioral input given. In our case we assume the user has already verified that the solution executes the behavior correctly by using a functional set of test vectors. One solution to testing chips is to create a set of test vectors to control and observe every fault possible in the chip design. This is called through the pins testing (McCluskey, 1986) . In other words we wish to detect at the output pins the presence or absence of faults while applying proper stimulants at the input pins. Another approach to test is called the structured design for test approach (Williams, 1983, McCluskey, 1986) . This approach, discussed in section 11.2.2, modifies the design to increase the testability or ease the generation of test vectors.

There are two main reasons why test is important. One is the cost view. The cost to detect an error increases by a factor of 10 at each level of design (Williams, 1983, Goel, 1980) . For example the chip design, board design, system design, and system design in the field are the four main levels where this cost factor increases, for example, from 10 at the chip level to 10,000 at the systems field level. Hence to avoid these large costs, testing at all design levels, from the chip to the system, is important. Testing at the chip level, the most complex of the three levels, due to the large number of faults and inability to probe internal nodes, requires detecting faults present at any of the internal nodes of the chip. Testing for faults on board-level wires and on connections between boards are required at the board and system level respectively. At these later two levels, the number of faults is smaller and most can be directly probed on board or at the system level interconnections between boards. We will concentrate our discussion in this book on chip level testing. The degree to which test is important may also depend upon the application. For example in military, avionics or automobile applications the test requirements may be very high.

Two related topics in test are redundancy and diagnosability. Redundancy at the systems level is very useful for fault tolerant design. However at the chip level redundancy leads to undetectable faults thus causing problems. We will not discuss system level redundancy. Determining why a chip is failing or where the fault exists is the purpose of a diagnosis test tool. These test tools maintain diagnostic libraries which relate output responses to faulty nodes. Diagnosis will not be discussed in this book.

The test stage is most traditionally viewed as occurring after the layout stage (Agrawal, 1984) . In particular for through-the-pins-testing the test vectors may be generated after layout and often this test generation process could continue until the fabricated chips are returned from the foundry. For structured design techniques the incorporation of scan registers would occur after the structural design is completed before layout. However the scan chain interconnection may occur after layout as in (Agrawal, 1984) . We could view the test process as a perpendicular line extending from the Y chart as shown in figure 1.3. In this figure the test stage can be performed after the system is laid out by arrow c. In structured test, transition a, b and c would occur. If test constraints are not met, the transition arrow, d, from the test axis to the structural axis illustrates the required redesign for testability. This may also occur if the area and time constraints are exceeded during the test stage.

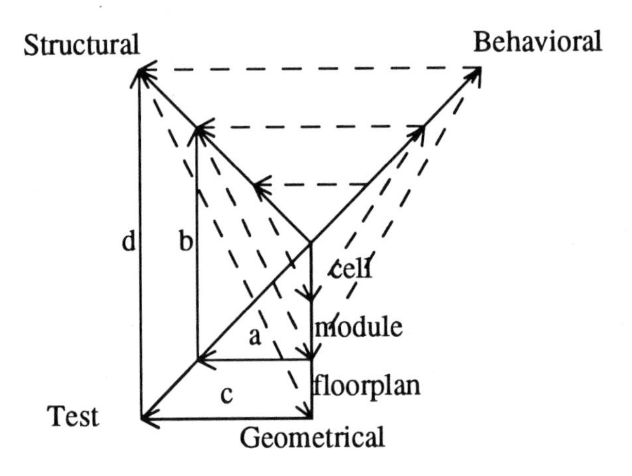

Figure 1.3. Modification of the Y chart to include the test stage.

The test problem is further complicated by the fact that as the number of transistors increase (from VLSI to ULSI) the number of primary inputs and outputs (or those accessible by pins of chip) do not proportionately increase (Tsui, 1986) . Thus it becomes even more difficult to access internal nodes of the chip for control and observation by using through the pins testing. The test generation problem for combinational circuits is NP-complete (Ibarra, 1975, Fujiwara, 1982, Aho, 1974) . The structured design for test problem, discussed in section 11.3, has also been shown to be NP-complete (Craig, 1988) . Approaches to the test problem are discussed in detail in sections 11.2 and 11.3. Some structured design for test techniques have been automated (Agrawal, 1984, Craig, 1988) , and some research such as (Abadir, 1985, Beausang, 1987) and (Fung, 1986, Gebotys, 1989) have discussed and published their integration of synthesis with test. These will be discussed in section 11.3 and 11.6.

1.5 BOTTLENECKS AND OPEN ISSUES

One high level bottleneck of the VLSI design cycle is the integration and synthesis of analog and digital behavioral specifications of a VLSI system. Different design methodologies and complex interfaces between the two domains pose many challenges for the design automation industry. There is a lack of DA tools to support these designs and currently industry relies on the communication between designers to define and design a correct interface. It has been estimated in the literature that it would take a larger effort than the design itself to take a self contained synchronous synthesized design and modify it to interface to other hybrid processes (Zahir, 1989) .

Since the design cycle steps are interdependent, low level bottlenecks can be partially alleviated by better high level design exploration. For example an architecture with fewer interconnect will decrease the problems at the lower level by easing the layout task to be perfomed. This is one example of relationship between the technology and the high level design.

In summary there exist a number of open problems in high level CAD for VLSI. The problems we will focus on are related to high level behavioral synthesis and are outlined below. In addition we believe these play a major role in currently preventing the high level tools from being accepted in industry.

1. Support for complex interfaces and timing constraints.
2. Optimized architectures for area-delay cost functions.
3. Long testing times for complex VLSI designs.

Problem one defines a realistic need to provide practical and usable tools for the mixed analog and digital or large systems VLSI design. Problem two delineates the requirement to make better decisions at the high level by providing DA tools which can communicate between different methodologies and make accurate estimates of the effect of high level decisions on the final systems design. Area (Sarma, 1990) and delay optimized architectures must be synthesized by these tools. The third problem defines a need to efficiently test a design to increase the probability that there exists no functional, logical, or performance errors in the fabricated chip. One possible solution to all these problems is to provide a rigorous adaptable mathematical framework (Gebotys, 1991x) which can support optimized design exploration. In addition it should model complex timing constraints and interfaces which may be combinations of of digital synchronous, asynchronous or analog processing units.

State of the art synthesizers to date can find at best "locally optimal" architectures with respect to an area delay cost function, and support simple timing constraints. Very few synthesizers have demonstrated how to use regularity and hierarchy of input algorithms to decrease the problem complexity. Hardly any architectural synthesizers at all even consider testability. We will focus on these problems in the next section and later in the text present a methodology to solve and advance the state of the art.

1.6 FOCUS OF TEXT

The aim of this text is to attack the three open high level CAD issues addressed in the previous section. Our solution is to provide a digital synchronous architectural synthesis tool which supports interfaces between different domains such as separately clocked synchronous processes, asynchronous circuits, and analog signal processing modules. The new general contributions of this research are outlined in the five points below.

1. Cost-constrained optimized high level VLSI architectural synthesis of digital synchronous systems.
2. Both local and complex timing constraints are to be supported for interfacing to asynchronous, analog or other external processes.
3. To provide a theoretical framework in which synthesis design automation tools can be developed for different types of architectures and clocking.
4. To explore a new mathematical approach to solving the synthesis problem. This approach involves a polyhedral approach aimed at providing global optimum solutions.
5. Ensure designs are testable at the architectural level through exploring design for test plans and structures.

We will not address other fields of synthesis such as logic (or controller) synthesis. Higher level behavioral partitioning techniques and transformations (such as those used in optimizing compilers) will also not be addressed, except to show how we can use their output for concurrent architectural synthesis whose inputs are partitioned behaviors or transformed code.

In this chapter we have looked at how problems in architectural synthesis impact higher level problems of system design and lower level problems such as routing. It was also outlined in general how improvements in architectural synthesis will improve the overall VLSI design cycle time. The next chapter will look closer at architectural synthesis with respect to its input and output primitives. The definition of input primitives for defining interfaces to external processes such as analog or asynchronous signal processing are also presented. Support for these interfaces are necessary in order for architectural synthesizers to have an impact on industry. It is our opinion that generally they have not received enough attention.

2.

BEHAVIORAL AND STRUCTURAL INTERFACES

This chapter will briefly discuss the structure of input and output primitives for high level architectural synthesis tools. The general structure of the behavioral input to an architectural synthesizer and a definition of its interface to external processes will follow below. Interface descriptions for analog and asynchronous or data dependent tasks are examined. Both the definition of a schedule and the specification of hardware primitives output from an architectural synthesizer are also included.

2.1 INPUT TO AN ARCHITECTURAL SYNTHESIZER

There have been many different languages and types of flow graphs constructed for describing behavioral input to previous architectural synthesizers, however we will not review these in detail

(Barbacci, 1981, Kuchcinski, 1988) . Unfortunately there are no standards for input languages of architectural synthesis. We will examine why a generalized directed acyclic graph (or generalized DAG) is in our opinion the most useful input representation for architectural synthesis even if it is not explicitly constructed. The DAG also serves as an important medium for describing the function of different architectural synthesis subtasks (McFarland, 1988) .

There are many different types of behavior (or input algorithms) ranging from matrix multiplication (and digital filters) to communication protocols. The representation of control in a behavioral description will also be outlined. The difference between controller synthesis and architectural synthesis is defined with respect to the input primitives, output primitives, and their mapping of software to hardware (driven by the implementation technology of output primitives).

Flow Graphs

Compilers have an intermediate form consisting of a mixture of flow graphs and DAGs. These intermediate forms play an important role in the efficient mapping of software to hardware especially for multiprocessor architectures. The DAGs in general provide an excellent medium for parallelism extraction and are used in conjunction with flow graphs in many optimizing compilers (Ellis, 1986) . The flow graphs, defined in compiler theory (Aho, 1974) , are used to define the control of a software specification. The nodes of the flow graph represent computations to be performed. These computations are essentially basic blocks of code, represented by DAGs. Basic block of code represent straight line code, which is code that contains no branch or loop constructs. An algorithm for constructing a DAG from straight line code is given in (Aho, 1974) . The arcs of the flow graph represent the flow of control. For example figure 2.1 illustrates the control and data flow graphs merged for representation of an input algorithm. The loop has an arc originating and terminating at the node representing the code inside the loop.

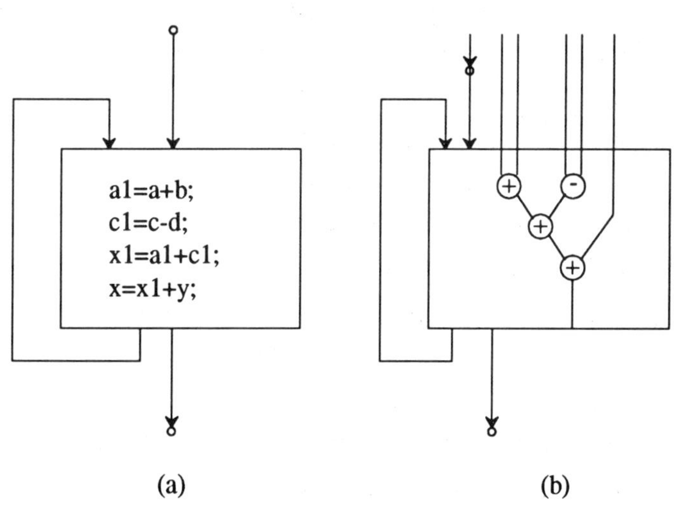

Figure 2.1. Merged DAG and flow graph for input representation for architectural synthesis.

In architectural synthesis many researchers have tried to merge the two graphs so there exists one medium with both control and data constructs. At a higher level there have been a number of languages introduced for input to architectural synthesis, such as ISPS (Barbacci, 1981) , Pascal (Kuchcinski, 1988) , and many others such as VHDL (which is also used as a standard for all levels of CAD representation). These languages have intermediate forms which bare some resemblance to DAGs, such as the value trace (Walker, 1987) .

In the following discussion we will demonstrate why DAGs in a flow graph can in fact represent more information than some higher level languages. Let us consider the following example of a matrix multiplication. In a mathematical notation it is: $Ac^T=b$ or $\sum_j a_{i,j}c_j, \forall i$. and in a algorithmic notation it is: *for* $(i=1,..,m)$ $\{b_0=0;$ *for* $(j=1,..,n)$ $\{b_j=b_{j-1}+a_{i,j}c_j\}$ $\}$. This can be represented as a number of different types of DAGs depending upon the order of operations. Each DAG may have significant differences in lower bounds on execution time. For example in (Papadimitriou, 1990) execution time is to be minimized and an infinite or very large number of processors are available. Therefore in their DAG, a tree is formed with multiplication operations at the leaves (degree one) and other nodes (of degree three) are the addition operations. A DAG is formed for each b_j calculation. This DAG is shown in figure 2.2b), where 9 clock periods are required for three multipliers and three adders. However in another application where accuracy of the computation is very important, the algorithm can be implemented as multiplier accumulator streams shown in figure 2.2a), requiring 8 clock periods for three adders and three multipliers. Each DAG may compute different values due to the ordering of the operations and error truncation. In this example and others the DAG offers the clearest representation for input to a high level architectural synthesizer as compared with languages that do not specify the order of operations. In summary we will focus on the mapping of the DAG to optimized hardware as opposed to the problem (ie. matrix multiplication) which may be represented by many different DAGs.

Instead of using a single language that many readers may not be familiar with we will instead use the generalized DAG, illustrated in figure 2.1, as the notation to represent an input primitive throughout this text. This avoids the ambiguity of operation ordering in languages and provides a good example for illustrating the architectural synthesis subtasks. In summary the notation uses control nodes (for branches, joins, etc), operation nodes, arcs for data transfer (and hence implied partial order),

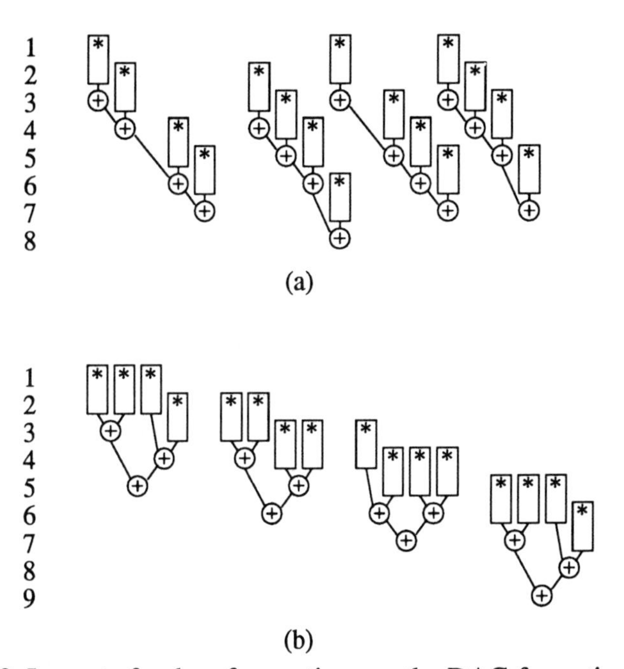

Figure 2.2. Impact of order of operations on the DAG formation.

arcs for partial order alone, and the specification of timing constraints.

2.2 INTERFACE PRIMITIVES FOR EXTERNAL PROCESSES

The definition of an interface is perhaps one of the most important features for behavioral input description for an architectural synthesizer, yet it has not been given enough attention in the architectural synthesis field. Previous researchers had assumed that the interface was not critical to the system performance and post processing was used to synthesize circuitry (Borriello, 1987) . Interfaces are very important since most custom chips are not designed as standalone systems. Very often a correct architecture or schedule cannot be guaranteed unless interface constraints are obeyed. In such cases, interface constraints may have a significant impact on the final speed or area of the chip. The complexity of an interface may vary from a simple data transfer off of a chip, to requesting data from a cache controller that is shared with other processes. In the later case the transfer of the requested data may occur after an unknown amount of time.

We will present four categories of interfaces and show how all other instances of interface constraints, that we know of, can be mapped into these categories. Secondly we will show that it is necessary to know or estimate the clock period of the design to be synthesized. The controller and architecture synthesized are responsible for transfer of data to and from interface circuitry at valid times in the most optimal manner. In the most optimal manner may mean to minimize the total execution time and therefore process the incoming data as soon as it arrives.

The Boundary

The four categories of interface constraints are (1) local, (2) analog, (3) asynchronous bounded and (4) asynchronous unbounded. Some local interface constraints are minimum, maximum or a combination of both timing constraints. The more complex constraints, (2) through (4), involve interfacing to analog or asynchronous processes. An example of the different layers of circuitry required to interface an external analog or asynchronous process with a digital synchronous process is shown in figure 2.3. The far left dashed line indicates the division of circuitry that we are concerned with. To the left of this line is the digital synchronous circuitry that will be synthesized. The multiplexor is used for illustrations purposes only. The first input to the multiplexor is an analog interface. An example of this interface is sampling an analog signal at a fixed rate and transforming this signal into a digital value (using an analog to digital converter circuit, ie. performing sample and hold (S&H) which outputs analog discrete time signals, and quantization (Q) which outputs digital discrete time signals) to enable subsequent digital signal processing. We will therefore define an *analog interface* to a synchronous digital circuit as a sequential synchronous data input or data output at a fixed rate. The third and fourth categories of interfaces are asynchronous interfaces. An *asynchronous interface* of a digital synchronous system is defined as inputs synchronized with the global clock but are indeterminate with respect to the control state of the system (Hayati, 1989) . In figure 2.3 a synchronizer (Synch.) is used to illustrate the possibility of having to synchronize the external signal with the clock of the digital component being synthesized. An example of an asynchronous interface is receiving data input from asynchronous circuitry or data dependent operations. Other examples of an asynchronous interface can result from transferring data between two synchronized processes, where each process uses a different local clock. It may also be possible that the two processes are using the same global clock, but the processes are loosely coupled causing the delay in clock signal to vary and therefore behave as if it were an asynchronous interface. Another example is the transfer of data (not necessarily at a fixed rate) to an analog process (DAC) for analog signal processing and subsequent receipt of the new analog processed output data (ADC). In this case we do not particularly care whether the external process is analog or digital. We can use an asynchronous interface if the processing time is data dependent or we can use minimum and maximum constraints otherwise.

Minimum and Maximum Delay Constraints

There exist many examples of minimum and maximum timing constraints. These constraints may also be applied locally to a DAG for example to describe a delay of two clock periods (or control states) for a multiplication operation. In this case a minimum timing constraint between the multiplication operation and the next operation which receives the output data can be used to represent extra clock periods

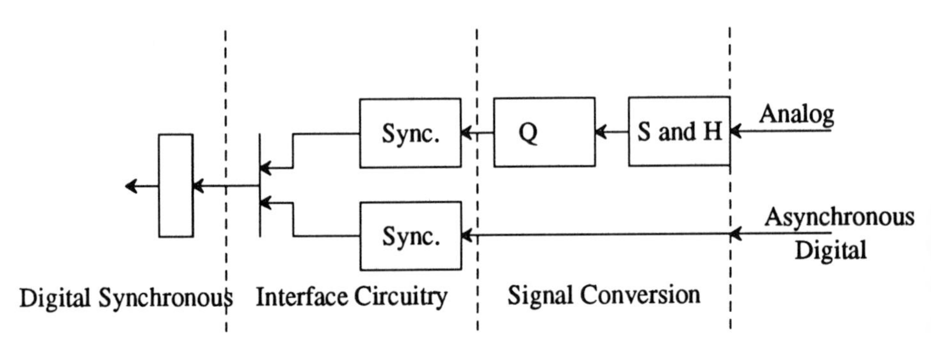

Figure 2.3. Levels of Interface between digital synchronous circuitry and external processes.

required by the multiplication to perform its function. In an interface scenario, the minimum and maximum timing constraints are also very important. For example if the output data must be valid in an output port register for at least three clock periods, in order for an external process to read the data, then a minimum timing constraint of three clock periods is required on the storage of the variable. On the other hand if incoming data is only valid at an input port for three clock periods then a maximum timing constraint of three clock periods on all operations which use this variable is required.

Analog Interface

There exist many examples of analog interfaces, as defined in this chapter. Let us assume that t_c is the period of the clock (in nanoseconds), and the incoming data, d_s, is arriving at a fixed rate of one sample every j_s clock periods (or every t_s nanoseconds), where $j_s = \lceil t_s / t_c \rceil$, $t_s \geq t_c$. If $t_s \leq t_c$ then we assume a high speed interface will collect the data into a large register (or queue) which is available to the synchronized system to be synthesized at each clock cycle. Other choices for the design with high speed interfaces will be discussed in chapter 5. Also let us assume that the digital synchronous behavior to be synthesized must ensure that all initial computations on the previous incoming data d_s have already been completed before the next data value arrives. Operations which input d_s must be scheduled after d_s arrives and before d_{s+1} arrives. Assuming the same computation is to be performed on each incoming data value then the algorithm would be a part of a loop, where at each iteration new data is received. Therefore a fixed timing constraint between operations which input d_i in successive iterations of i should be equal to the j_s.

Asynchronous Interfaces

We will now study the impact of asynchronous interfaces on digital architectural synthesis. As discussed in the previous section this is not necessarily an interface with asynchronous circuitry , but may also include interfaces to data dependent processes (Ku, 1989a) . Two types of asynchronous interfaces are discussed below, bounded and unbounded. We will show that the later case can be transformed into a bounded and wait-state interface. These interfaces are quite complex and impact both the scheduling of the DAG and the allocation of hardware. We will further discuss these interactions in chapter 7.

Bounded Delays.

A bounded asynchronous interface is defined as an asynchronous interface where the lower bounds and upper bounds on the indeterminate control state are known. The bounded asynchronous interface can be represented by constructing a flow graph from the DAG with the asynchronous interface. In figure 2.4 a) the DAG is transformed into the flow graph of 2.4b) where the bound of three clock periods for receiving data from K_a is represented as a three way branch, starting after basic block (defined in section 2.1) A, with a delay of one cstep for each branch. Before the three way branch is placed in the DAG the operations must be partitioned into interface dependent and interface independent basic blocks. In the first case an operation is *interface dependent* if there exists a path from the vertex representing the asynchronous operation to the vertex of the specific code operation (ID and C are interface dependent in figure 2.4). If there is no path between these two operations in the DAG then the operation is interface independent (such as the basic block of code B in figure 2.4). The flow graph is constructed by placing interface dependent operations in a separate basic block. The interface independent basic block may eventually end, at a particular control state, after which all code operations must precede the interface dependent code (in figure 2.4 basic block C is all interface dependent code).

Unbounded Delays.

Unbounded or ∞ bounded asynchronous interfaces are asynchronous interfaces where the bound on the control states is not known; for example designing with a synchronizer, or data dependent loops. An example of a general DAG with unbounded delays can be shown in figure 2.4c). Although most researchers discuss partitioning the graph at the ∞ bounded operation vertices, we will discuss a different partitioning into three groups of operations for DAGs where the interface independent basic block ends before the interface dependent operations as shown in figure 2.4b). In these cases the ∞ bound can be removed and it is possible to decompose the unbounded interface into a bounded interface and a wait state. The bounded interface occurs from the earliest control state that the asynchronous operation may output data to the last control state required to complete the interface independent code. In figure 2.4c) this requires the three way branch. After this cstep, if data is still not available from the external process, the controller must essentially wait (ie.

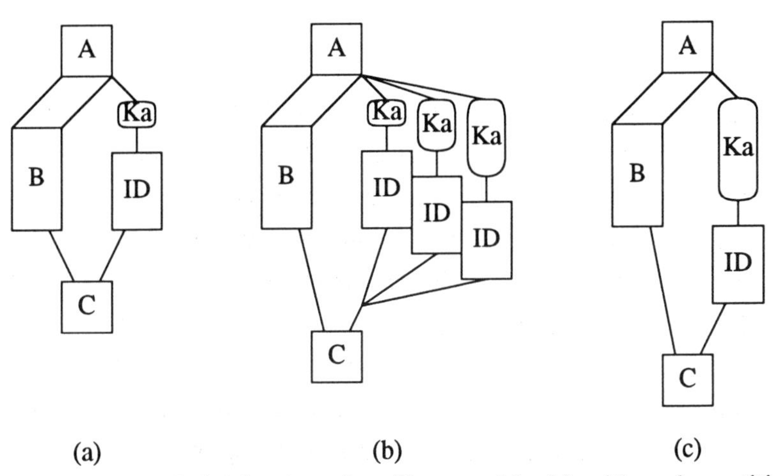

Figure 2.4. Bounded delay interface illustrated in (a) with code partitioning. In (b) the conditional branches are used to model this bounded delay and the wait state in (c) along with (b) are used to represent an ∞ bounded interface.

perform no operations until the data is available from the asynchronous operation). In figure 2.4c) the schedule and allocation of DAGs with bounded or unbounded interfaces can have a significant impact on controller complexity and area and delay of the architecture.

2.3 OUTPUT PRIMITIVES FROM AN ARCHITECTURAL SYNTHESIZER

In this section we will discuss hardware architectural primitives that are used in architectural synthesis. Essentially the hardware primitives are output from the synthesizer and later refined into more detailed modules to be placed and routed. The output primitives are divided into storage primitives, processing elements, and interconnect primitives. Each section below defines the generalized primitives which are necessary to understand architectural synthesis.

Processing Elements

The processing elements receive input data and after a defined amount of time produce new output data. The processing elements perform computations which we will call functions. This terminology is used to avoid the confusion with the term operations, used to describe the nodes of the DAG or specification of the input algorithm. For example an adder (a processing element) performs the function addition. Some processing elements may have more than one function they can compute. The term *functional unit* is used to refer to a particular processing element and each functional unit has a corresponding module in the VLSI library of cells. Therefore each functional unit is defined by a set of functions that it can perform. The set of functions of two functional units may or may not overlap. Additionally two functional units may have identical sets of functions but they may require different amounts of time to compute their outputs. To distinguish functional units by these characteristics we use the term, *type*. The *type* of a functional unit is the most detail we will use for high level synthesis. For example one type of functional unit is a two cycle multiplier and another is a pipelined multiplier. Both functional units compute the same function, however, their timing characteristics are different. An ALU and an adder are two other types of functional units. For convenience we will illustrate the functional units using a circle or vertically placed rectangle.

Storage Primitives

There are many different types of storage primitives. In fact most systems have a hierarchy of storage starting at the bottom level with registers, register files, and moving up to memory caches, main memory, etc. The simplest and most common storage primitives for architectural synthesis are the register and register files. Their difference is illustrated in figure 2.5. We will concentrate on *registers*. In phase one of the clock a master/slave register transfers the input data to the output of the register. In other words, a new data value is placed at the inputs of the functional unit. During phase two, the output from the functional unit is latched into the register. Note that the bus is only active during phase two. The register files can be visualized as splitting a register into two latches and moving one latch to a register file and keeping the other latch at the input of the functional unit. When this is done the busses can be used for data transfer during both phases of the clock. Transfer from the register file to the latch at the input of the functional unit occurs at phase one. Phase two transfers the output data from the functional unit onto the bus and into the register file. Phase one in the register file architecture is one example of a storage to storage primitive transfer. It is interesting to note that by using only one bus for input and output from each register file as in (Haroun, 1989) some variables must be stored in more than one register file for concurrent accesses with other variables. This can account for more latches in a register file architecture than registers in a register architecture. We will assume that we are dealing with the register architecture, shown in (a), unless otherwise stated that the architecture is the register file architecture. For illustration purposes registers will be represented by a horizontally placed rectangle with a horizontal line through it to represent the two phase operation.

Interconnect Primitives

Interconnect primitives are an important part of architectural synthesis, and often the most controversial. Interconnect primitives can include busses, multiplexors, demultiplexors, and multi-level combinations of these interconnects. It is not clear how to measure interconnect so that high level architectural solutions (that have not been placed and routed) can be compared. In the most general terms we will define a

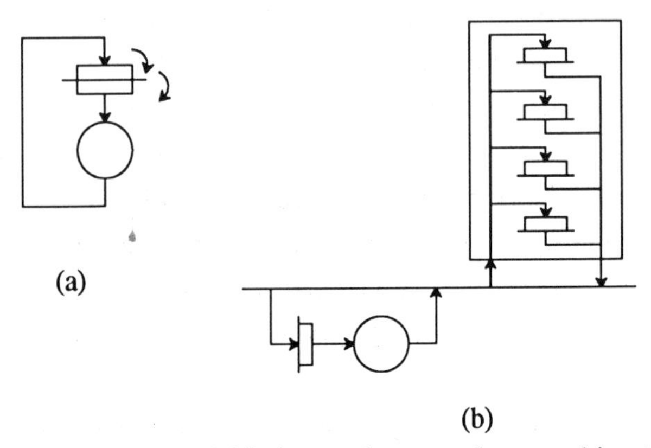

Figure 2.5. An example of (a) the random topology architecture and (b) the register file architecture.

general *bus* as an interconnection with one or more inputs and one or more outputs. This is a good definition to use since it provides an exact number of general busses at a high level of abstraction, unlike other measures of busses (Haroun, 1989) which do not account for multiplexors at the inputs to registers or functional units. These busses interconnect storage elements to functional units, and vice versa. For example in figure 2.6a) there are 3 busses. The first bus connects registers a and c to functional unit f, the second bus connects register a and b to functional unit f, and finally the third bus connects functional unit f to register b in figure 2.6a). These three busses are also shown in figure 2.6c). In figure 2.6b) there are 2 busses and 1 multiplexor. The two busses are illustrated with horizontal lines and the multiplexor connects both busses to one input latch of functional unit f in figure 2.6b). Using the general bus definition, the architecture in figure 2.6b) also has three busses. This is obtained by adding the number of busses (connecting functional units to register files) to the number of multiplexors (connecting register files to inputs of functional units). Other measures of complexity have been proposed as counting the number of bus drivers or multiplexor inputs, number of equivalent two to one multiplexors (Ly, 1990) or even the number of connections to busses or multiplexors (Cloutier, 1990) . In general there is no standard for comparing interconnects even for solutions using the same type of architecture, for example register architectures. It is clear that for a schedule and an allocation of hardware resources (prior to binding) the only measure one can obtain is the number of general busses. Additionally interconnect primitives, which will not be explicitly analyzed in this manuscript are bus connections or bus drivers, since these are defined during binding.

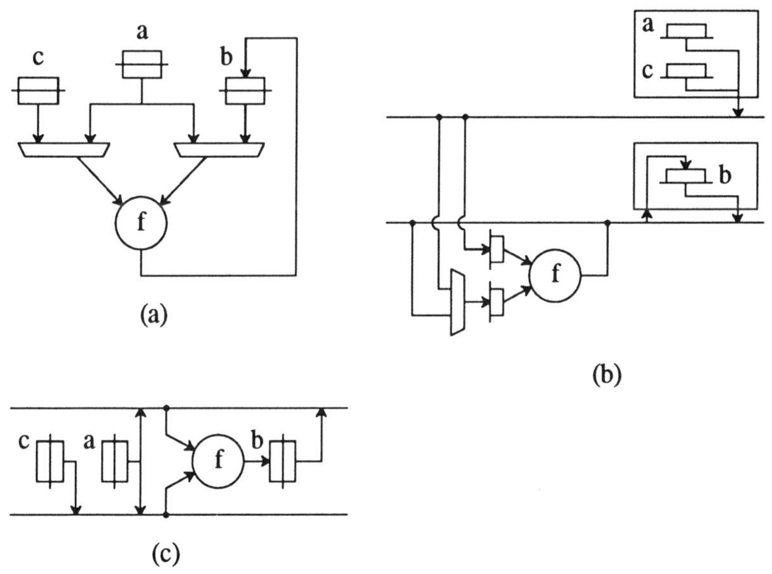

Figure 2.6. A behavior implemented in (a) random topology and (b) register file architecture, and (c) a random topology with busses instead of multiplexors.

In this chapter we have defined specifically the input/output and interface specifications associated with architectural synthesis. Now that the reader has an idea of what goes in and comes out of an architectural synthesizer we can define the transformations that must be performed in the next chapter. Chapter 3 also gives a brief introduction to previous research in architectural synthesis, concentrating on optimization approaches as opposed to the many heuristics that have been developed. Different approachs to solving architectural synthesis are examined from independent subtask optimizations to more recent simultaneous approaches are covered.

PART II : REVIEW AND BACKGROUND

3.

STATE OF THE ART SYNTHESIS

State of the art high level synthesis approaches will be reviewed in this chapter. Each section will provide a definition of the problems and an introduction to the mathematics involved in solving these problems. We examine previous research as it relates to each problem including independent subtask optimizations, simultaneous approaches to synthesis, and mathematical models. In addition we will briefly discuss feasibility models, cost functions, high level partitioning tools and timing considerations in logic and architectural synthesis.

3.1 TERMINOLOGY AND SUBTASK DEFINITIONS

We define some frequently used terms the reader will find helpful in understanding the function of different subtasks of synthesis research. As briefly discussed in section 2.1 there exist various media for input representation. We will assume the most general (intermediate) form of an input algorithm, a directed acyclic graph (DAG), where the nodes represent the code operations, and the directed edges (arcs) represent the variable transfers between code operations. Any algorithm or z-diagram can be represented by a DAG. Hardware output primitives for architectural synthesis were defined in chapter 2. However we will review briefly some additional terminology here. *Modules* refer to hardware units which will be defined (in functionality) with operations at some later point. *Functional units* refer to digital hardware units (for example an ALU) that perform a defined set of computations on the input data and provide new output data. For example one functional unit may be a 2 cycle pipelined multiplier and another functional unit may be a 3 cycle non-pipelined multiplier. *Scheduling* refers to the assignment of code operations to time. Since processing is synchronized with a global clock, time is an integer value. We use the term control step (cstep) to represent the state of the synthesized architecture where control step 1 is present after the architecture is powered up and initialized. The execution time of the algorithm (Te) is defined as the minimum number of csteps required to execute the input algorithm or DAG on the synthesized architecture. *Allocation* is the determination of the number of hardware units such as functional units, registers, and busses. For example, four registers may be allocated, however the variables that are stored in each register have not yet been determined. A schedule may require 3 modules, which may be defined (through binding code operations; addition and multiplication) as 2 adders and one add/multiply functional unit. If the add/multiply functional unit does not exist in the library then 4 functional units (3 adders and 1 multiplier) may be necessary. The number of modules is a lower bound on the number of functional units to be allocated. In general the term resource will refer to functional units, busses, and registers.

Additional terms will be used to compare with other synthesizer techniques. When one fixes a number of resources, for example one can fix the number of registers at ten, this means that one does not minimize the number of registers, but places an upper bound on the number of registers of ten. *Estimated* refers to using some heuristic to minimize a particular resource. *Calculated* refers to an exact computation of a number of units after a schedule is found. A fixed schedule or hardware allocation means that the schedule or allocation has already been performed by some earlier algorithm and therefore is a constraint on the remaining problem.

The output of the architectural synthesizer that we will address are the following:

- total number of control steps, functional units, busses and/or multiplexors, registers and/or register files, memory.
- *scheduling*: code operations to control steps.
- *functional unit allocation and selection*
- *register allocation*
- *interconnect allocation*

The hardware units listed above were defined in section 2.3. One must also determine the type of hardware (type of functional units or memory versus registers) to be used in the final architecture. In some cases the former is done during architectural synthesis. The final schedule and binding produced by the architectural synthesizer can be transformed into a control table for input to a logic synthesizer.

The architectural synthesis problem involves many subtasks such as scheduling (S), resource allocation (A), and resource binding (B). We will use the term resource to describe the hardware primitives or registers, functional units, and busses. However each of these steps are heavily interdependent. An example of the interdependence between the subtasks is shown in figure 3.1. For example a fixed schedule directly determines the minimum number of functional units and registers (allocation). The subsequent binding of these resources directly determines the minimum number of multiplexors (allocation) required in a multiplexed architecture (S $\Rightarrow$ A $\Leftrightarrow$ B). Another design approach which illustrates the interdependence in figure 3.1 is to first perform resource allocation. This allocation will constrain the scheduling and subsequently constrain the binding (A $\Rightarrow$ S $\Rightarrow$ B). It is also easy to see that binding affects scheduling (B $\Rightarrow$ S). For example operations bound to the same resource cannot be scheduled at the same time.

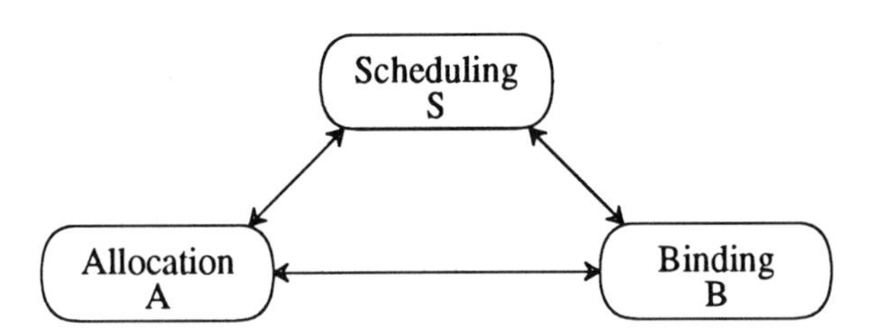

Figure 3.1. Subtask interdependence in architectural synthesis.

Ideally the optimal approach to solving architectural synthesis is to simultaneously consider all tasks at the same time. However since this is a very complex approach most researchers have concentrated on one or a limited number of subtasks to be solved simultaneously. We will briefly review research in this field with emphasis on graph theoretical results and integer programming (IP) approaches. We will describe the complexity of these subtasks and overview the iterative/simultaneous approaches to architectural synthesis. More detailed analysis of architectural synthesis material can be found in papers such as (McFarland, 1988, McFarland, 1990) .

3.2 HIGH LEVEL TRANSFORMATIONS

Many high level transformations of algorithms for architectural synthesis have been borrowed from compiler research (Aho, 1974) We will concentrate on reviewing partitioning of behaviors and some (compiler-like) transformations used within the context of architectural synthesis.

Partitioning of behavioral descriptions for architectural synthesis of multichips has been investigated by a number of different researchers. APARTY (Lagnese, 1989) evaluates different partitions of the DAG but schedules the graph entirely without partitions. The algorithm passes heuristic suggestions to other allocation and binding subtasks to use partitioning information only during allocation. This occurs because it is not known how to define interfaces between the partitioned behavioral specifications in order that they be concurrently and independently synthesized. Also in (Gupta, 1990) partitioning of behavioral specifications is performed after a binding (which defines hardware sharing) for multichip design. The hypergraph partitioning bounds the latency of the partitioned implementation.

Other research (Depuydt, 1990) deals with partitioning large complex signal flow graphs. Various clustering techniques are used to partition into more manageable sized flow graphs for separate scheduling using better or more optimal techniques which work well on smaller input flow graphs.

Research at Carnegie Mellon University (Walker, 1987) has examined implementation of behavioral code transformations in a user interface environment that is tied into their architectural synthesizers, DAA. Although these transformations are the same as those found in optimizing compilers [1] (Ellis, 1986) , it is unknown what effect the transformations have on the final architectural synthesized design.

1 Optimizing compilers do not optimize, but they heuristically attempt to extract further parallelism from the input code.

Flamel (Trickey, 1987) used an algorithm which performed many basic block [2] transformations to increase the parallelism of the input algorithm and then subsequently synthesize the architecture. Different types of merging basic blocks and unrolling loops were performed. The new transformed input algorithm was then synthesized using an integrated scheduler and folding technique. This technique for increasing parallelism showed improved performance by implementing programs that would run 22-200 times faster than a M68000 running the same program. No hardware sharing of mutually exclusive code was performed. Although we will not directly address high level transformations in this text, some will be used in chapter 10 and 11 with respect to future research in global optimization of synthesis.

Design Style and Clock Speed Selection

The clock speed and design style selection are interdependent. Design style defined in (Haroun, 1989) refers to the types of functional units, for example an adder or an ALU, to be used in synthesis. For example if one chooses a 115ns clock period and one type of multiplier with a 100ns propagation delay and 20ns delay adder, then one cycle is required by the multiplier and one cycle by the adder. However if the clock period is 130ns then it is possible to chain the multiplier and adder together, therefore defining a new type of functional unit (which can compute (x * a + b) in one clock period). Most DA systems assume that the clock period is defined before synthesis so that the operational characteristics of the functional unit are known. In fact after synthesis a finer grain selection of functional units can be performed to possibly further improve the design.

---

2 Basic blocks are sections of code, called straight line code, that contain no branches or loops.

3.3 INDEPENDENT SUBTASK OPTIMIZATIONS

In this section we will study the various subtasks associated with architectural synthesis. The graph theoretical problems, their complexity and solutions are discussed for independent and simultaneous solutions of subtasks. In chapter 4 we will outline the analogous integer programming representations of some of these algorithms, and further show the advantages of using integer programming formulations in chapter 6 and 7 for simultaneously solving more than one subtask and incorporating complex constraints.

3.3.1 Scheduling

The scheduling of a DAG without resource constraints can easily be performed in polynomial time (Foulds, 1981) using the well known critical path method (CPM). This algorithm calculates the critical path and the as soon as possible (asap) and as late as possible (alap) control steps (Foulds, 1981) for each node of the DAG. This algorithm executes in $O(n^2)$, where n is the number of nodes in the DAG. An example DAG, representing the operations $w{=}y{*}z;x{=}((a{+}b){+}c{*}d{+}w)$ and illustrating the asap and alap schedules, are shown in figure 3.2 a) through c). The bottom empty circle is used to ensure that the variables x and w are output at the end of the algorithm. The alap schedule can be calculated for any upper bound on the number of clock periods by incrementing the previous alap csteps by $(T_e^{UB}-T_{C.P.})$ number of csteps, where T_{CP} stands for the minimum number of csteps in the critical path. The asap schedule obviously is valid for any upper bound on T_e. Therefore this processing needs to be done only once per application (or input algorithm).

The asap and alap schedule have not been used for subsequent resource allocation in architectural synthesis with much success because they do not always produce designs with an optimal number of resources. In figure 3.2 the asap and alap requires 3 modules (2 * and 1 +) and 2

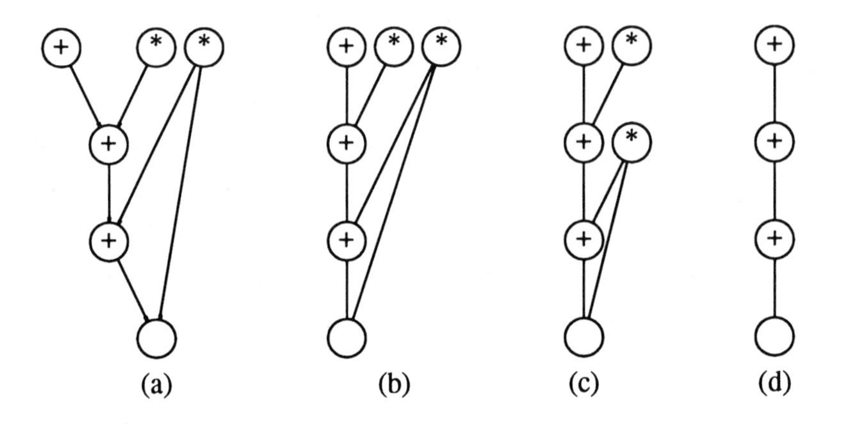

Figure 3.2. DAG (a) and corresponding asap (b), alap (c), and critical path identification (d).

modules (1 * and 1 +) respectively. However these schedules are very important for an initial analysis of the synthesis problem by providing the range of valid control steps (which do not violate any partial order constraints) for each code operation.

3.3.2 Resource Allocation

Almost all resource allocation in architectural synthesis problems for a fixed schedule have similar structure. We will represent a graph, $G=(V,E)$ as a set of vertices $\varepsilon\, V$ and edges $\varepsilon\, E$. In general the scheduled DAG is transformed into another (conflict, or compatability) graph. By further classifying this graph (chordal, interval) one can either solve the problem optimally using a known polynomial time algorithm or heuristically using a similar algorithm. We will use register allocation as an example to illustrate the transformation and solution process that previous research (Tseng, 1986) has examined. Not only is register allocation an interesting subtask, but as it will be further discussed in chapter 3.5 , its simple solution for basic blocks presented in this section becomes even more difficult (NP-complete) to solve simultaneously with the scheduling problem.

Although for general graphs some of the problems, such as vertex coloring, presented in this section are NP-complete, they can be optimally solved using known algorithms in polynomial times if the graph is of a particular type (Golumbic, 1980) . It is interesting to note that the same types of characterizations exist in integer programming (IP) and often for the same problems. We will discuss integer programming aspects further in chapter 4.

We assume that the DAG is scheduled in figure 3.3a) in four control steps (including the last cstep for the last node whose incident edges are the output variables). Each variable can be represented in an interval representation shown next to the DAG. In the interval representation, the lifetime of each variable is represented by a vertical edge starting at the cstep the variable is defined (output by a code operation) and ending at one cstep before the latest cstep where an operation uses the variable as input. This interval representation is convenient for register allocation because we have to find sets of variables, such that in each set the life-times of the variables are disjoint (or in other words no two lifetimes of the same set have the same cstep). Thus each set represents a register. We will next define the graphs and then define the algorithms.

The compatibility graph, G^c, is formed from the interval representation. Each edge of the interval representation becomes a vertex of the graph G^c. Edges are formed between all pairs of vertices in G^c whose corresponding variable lifetimes are disjoint (originally called "comparable" vertices (Hashimoto, 1971)). In other words two variable lifetimes are disjoint if there exists no cstep where the lifetime of both variables intersect. The conflict or interval graph (Golumbic, 1980) , G^i, uses the

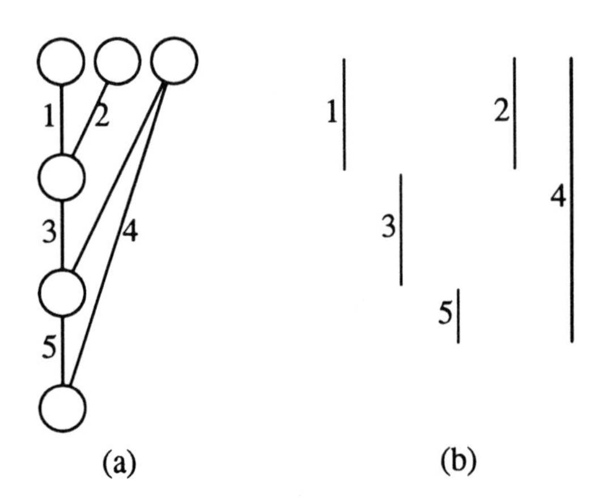

Figure 3.3. Scheduled DAG (a) and the variable lifetimes shown with an interval representation.

same definition of vertices as G^c however edges are formed between all pairs of vertices whose variable lifetimes are not disjoint or in other words have overlapping lifetimes (or are "incomparable"). Another characteristic we can observe from these two graphs is that G^C is the complement [1] of G^i.

Register allocation is performed on G^c by a clique partitioning algorithm. Clique partitioning essentially removes edges from G^c so that the remaining graph is a number of disconnected cliques. The algorithm tries to produce a minimum number of disconnected cliques. A clique of a graph G is a maximal complete subgraph. We will use the notation K_x to represent a clique on x nodes. For example in figure 3.4 there can be 3, 4 or 5 cliques in a partition. For the minimum number of 3 cliques there

1 The complement of graph G is $\overline{G}$; ($\overline{\overline{G}}=G$).

are two different possible partitions that may be used. The number of cliques is equivalent to the number of registers.

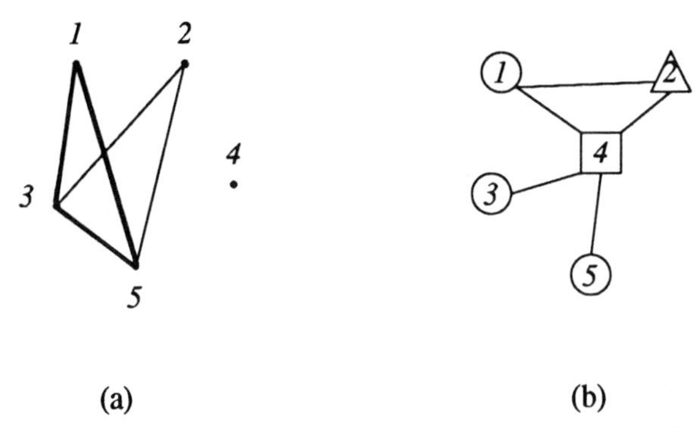

Figure 3.4. The compatibility (a) and interval graph (b) derived from the interval representation of figure 3.3

Alternatively the register allocation problem can be solved on graph G^i using vertex coloring. The vertex coloring of the interval graph, can be solved using a polynomial run time algorithm or the left edge algorithm also presented for solving channel routing problems in (Hashimoto, 1971) . The number of colors is equivalent to the number of registers. In fact the minimum number of cliques in G^C is equivalent to the minimum number of colors (or independent sets which cover the graph) in G^V. These two algorithms are hence complementary.

The clique partitioning approach was first presented in Facet (Tseng, 1986) . It was shown in (Springer, 1990) that a larger problems could be solved faster than using the interval graphs.

In the presence of conditional code there may be more than one edge used to represent a variable's lifetime. For example a variable defined before a branch on conditional code, but whose last use is at different csteps inside each branch. Thus the graph is no longer an interval graph and one cannot minimize registers in general. REAL (Kurdahi, 1987) heuristically extended the left edge algorithm for conditional resource sharing register allocation. However in (Springer, 1990) specific types of conditional code that formed chordal graphs (of which interval graphs are a subset), were identified thus showing that one could for some cases minimize the number of registers in the presence of conditionals. Minimizing registers in loops, where variable lifetimes are defined on a circle, was also solved by using an arc coloring algorithm in (Haroun, 1989) .

Functional unit Allocation and Bus Allocation.

Functional unit allocation is complicated by the fact that the mapping of operations to type of functional units may be a one to many mapping. In other words a selection of types of functional units for each operation must be performed. Many synthesis systems reduce this complexity to a one to one mapping, by preselecting the types of functional units, and therefore do not simultaneously select functional units when performing allocation. Facet (Tseng, 1986) performs functional unit allocation also using the clique partitioning algorithm. The user provides a scheduled DAG and Facet solves each allocation task, including register, functional unit and interconnect allocation, independently using a clique partitioning heuristic algorithm.

MIMOLA (Marwedel, 1986) uses a integer linear programming model (IP), with branch and bound solver, to obtain the number of functional units required for a fixed schedule. However it could not apply this IP to bind operations to functional units due to its large model size.

The problem of bus allocation with a fixed schedule is also very similar to register and functional unit allocation and busses are allocated after these allocations. The number of data transfers per cstep are used to calculate the number of busses. If one wants to allocated all general busses (multiplexors and busses) there is a problem with using global data broadcasts. A global data broadcast is a transfer of one data value from one source to more than one destination. If one counts the number of distinct sources (accounting for a global data broadcast as one transfer) then this will not account for extra multiplexors which may be required at the inputs of functional units. On the other hand if one counts the data broadcast by the number of destinations then one may overestimate for the number of busses. In most synthesis systems it is assumed that the extra multiplexors required will be substituted later in the design process, and the number of sources for data transfers is counted. Interconnect optimization with a fixed schedule and a fixed number of functional units, (Stok, 1989) for register-transfer file architectures with separate read and write clock phases was examined using a simulated annealing approach.

3.4 ITERATIVE AND SIMULTANEOUS APPROACHES

Scheduling and functional unit allocation were the first two most common subtasks to be considered simultaneously. Previous research (Garey, 1979) for scheduling multiprocessor systems such as list scheduling (Coffman, 1976) has had a large impact on the architectural synthesis application. We will use this application to introduce and define the problem. A brief overview the architectural synthesis applications will then be performed. This scheduling and functional unit allocation problem is similar to the precedence constrained scheduling problem formally defined in (Garey, 1979) as:

> " A set T of 'tasks' (each assumed to have 'length' 1), a partial order $< \bullet$ on T, a number of 'processors' and an overall 'deadline' $D \varepsilon Z^+$.
>
> Is there a 'schedule' $\sigma: T \rightarrow \{0,1,..,D\}$ such that, for each $i \varepsilon \{0,1,...,D\}$, $|\{\ t \varepsilon T : \sigma(t) = i\ \}| \leq m$, and such that, whenever t $<\bullet$ t', then $\sigma(t) < \sigma(t')$?"

This problem was proved (Ullman, 1975) to be NP-complete. The precedence constrained scheduling problem for DAGs with an intree structure (Brucker, 1977) were shown to have a polynomial time solution and outtree examples, both illustrated in figure 3.5, were shown to be NP-complete. This research was the start of a technique called list scheduling (Coffman, 1976) which has since been refined for architectural synthesis, such as (Pangrle, 1987, Paulin, 1989) .

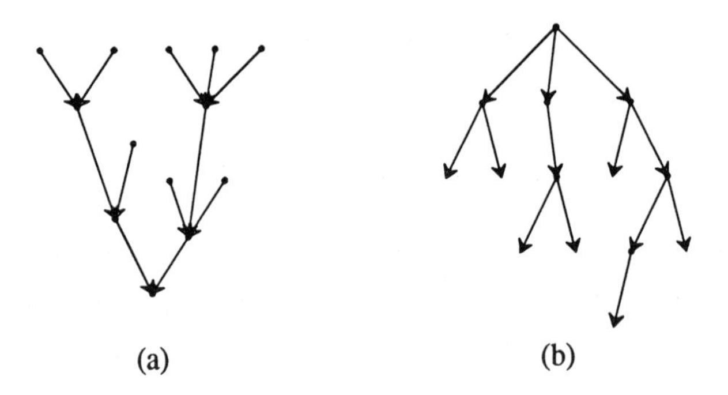

Figure 3.5. Intree and outtree DAGs, which are schedulable in polynomial and exponential time.

In general models for previous scheduling research consisted of a variable t_j where t_j's value is the time that job j is scheduled. Various objectives and additional variables representing release time, job time, or delivery time were used. These types of scheduling problems have been extensively studied in the literature and are an ongoing research topic (Hal, 1990) .

The partial order of the quoted precedence constraint scheduling problem represents a data transfer in the architectural synthesis model. The partial orders can be also represented by arcs in a directed acyclic graph representation of the set of tasks. The extensions to the formal scheduling problem for architectural synthesis include : limited mapping of tasks to processors; timing constraints; and complex task operation such as multicycled or pipelined processors.

Special Case Solutions

Research in mapping algorithms onto multiprocessor structures also examines the precedence constrained scheduling problem (Garey, 1979) . For an infinite number of processors one can schedule a DAG to minimize the makespan or execution time of the algorithm. In multiprocessor applications the assumption is made that each processing node of the DAG requires negligible time compared to the time for communication between processors. Therefore the problem in this research area is modeled as a function of the number of communication delays required to perform the algorithm (Papadimitriou, 1990) . Other research has shown that if we limit our architecture to two modules then given any DAG we can calculate the minimum execution time (Lawler, 1976) . This problem maps into a matching problem in a graph which is the complement of the DAG. The matching problem is to maximize $|M|$, where $M \subset E$ of a graph, $G=(V,E)$, such that each vertex is incident to at most one edge εM. An example shown in figure 3.6 illustrates a matching, $|M| =2$, thus providing an optimal schedule of 3 control steps for a 2-processor implementation of the five code operations (a,b,c,d,e). In fact a valid schedule could also be obtained using the matching algorithm.

If we increase the number of modules beyond 2 the problem is again NP-complete, since we are then looking for a restricted set of cliques of size less than or equal to the number of modules (>2). It is however interesting to look at this application since it illustrates the limitations of purely graph theoretical approaches to solving complex problems. For example as new complex constraints arise during the design cycle using purely graph theoretical approaches may not be viable due to the difficulty in adjusting these algorithms to the new constraints. We will now briefly review previous research that tries to simultaneously schedule and solve functional allocation tasks.

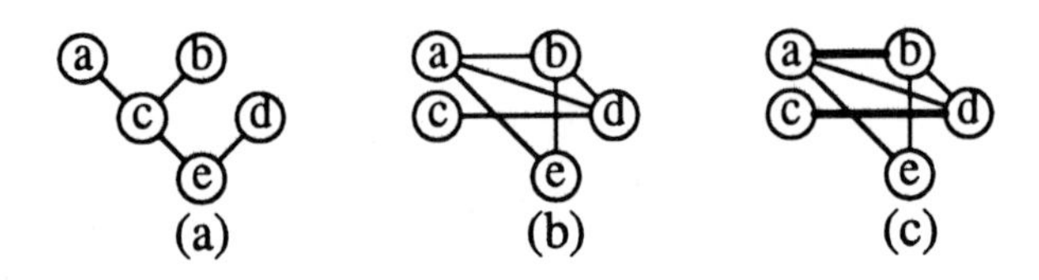

Figure 3.6. Illustration of restricted optimal scheduling for two modules

Previous Scheduling Research for Architectural Synthesis

Variations of list scheduling techniques are very popular in architectural synthesis as well as multiprocessor compiler design (Sarkar, 1989) . In general one fixes the number of functional units and then schedules operations in a prioritized order. The priority is set by the (alap - asap) value, where a smaller value has a larger priority. Operations are placed in a cstep based upon this priority until all functional units are exhausted. Then operations are placed in the next csteps in the same manner. HAL (Paulin, 1989) uses an iterative refinement heuristic algorithm based on force directed list scheduling to perform scheduling and functional unit allocation. Recently extensions to provide heuristics to minimize registers and interconnect have been incorporated. The number of parallel data transfers, using transfers with distinct sources counting as one transfer, were used to heuristically approximate the number of busses. However the exact relationship to number of busses was not defined.

3.5 MATHEMATICAL APPROACHES

The mathematical approaches to simultaneously solving more than one subtask of the architectural synthesis problem will be outlined in this section. In these examples the scheduling was simultaneously solved with more than one subtask. However no previous research to our knowledge has tried to simultaneously schedule and allocate busses, only estimates of busses are used to guide the scheduling task. These examples show how the previously studied independent subtasks, such as register allocation for a fixed schedule, now become very difficult to solve simultaneously with the scheduling subtask.

3.5.1 Branch and Bound

A MILP model in (Hafer, 1983) , solves simultaneous scheduling, functional unit and register allocation using a MILP (mixed-integer LP) formulation. In addition scheduling is done in real time and both registers and functional units are selected from a library. A nonlinear model was first formed and then linearized by the addition of binary variables. Unfortunately only very small examples could be solved due to the size of the model and the inefficiencies of the branch and bound technique. For example an input algorithm with 4 code operations required 87 variables, of which 46 had to be integers.

One of the first IP models for resource constrained scheduling was presented in (Baker, 1974) . This same model was recently used in a two step methodology in (Lee, 1989) . The IP formulation was solved using a branch and bound algorithm to produce a schedule that minimizes the number of functional units in one step and the sum of the lifetimes of the variables of the DAG (which heuristically minimizes the execution time and in some instances the number of registers) in the second step. Figure 3.7 shows an example where this heuristic fails to minimize the number of registers. Very fast execution times were obtained most likely due to the improved computer technologies available today as compared to 20 years ago. More importantly by using this two step methodology bounds are kept small by incrementally moving across the design space. However the bounding argument (which sets the previously solved number of functional units as an upper bound for the present optimization with a larger execution time possible) does not necessarily hold in all cases. For example very often as the execution time (or number of control steps) is increased the number of adders may increase at the added benefit of decreasing a more expensive functional unit such as a multiplier. These tight bounds, as will be discussed in chapter 3, are very important for solving any IP and in particular for branch and bound techniques (they greatly improve the performance). The model was later extended for functional pipelining in (Huang, 1990) and a heuristic partitioning strategy to decrease the size of the input algorithm, however register allocation could still not be incorporated.

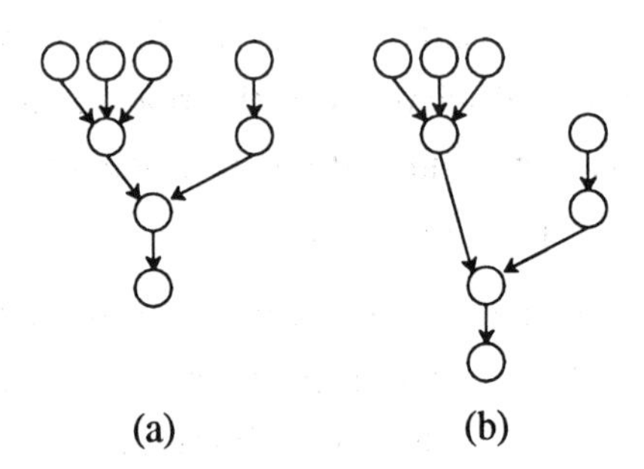

Figure 3.7. An example where sum of the lifetimes of the variables (sum) in the DAG does not decrease the number of registers but favors minimum execution time. In (a) Te=4 (minimum), sum = 7, 4 registers, and in (b) Te=5, sum = 8, 3 registers are required.

3.5.2 Simulated Annealing

A simulated annealing technique presented in (Devadas, 1989) solves simultaneous scheduling, functional unit allocation, and register minimization. The formulation includes a calculated number of registers, and an estimate of interconnect in its cost function. Since the cost functions are used to evaluate two dimensional placements (or fixed schedule and functional unit allocation), the number of registers could be calculated using the left edge algorithm. The number of parallel data transfers was used as a heuristic estimate of the number of busses as defined in HAL, however again the relationship was not defined. Another part of the cost function was called links, which tried to estimate the number of bus drivers or multiplexor inputs required. Both fast simple and slower more accurate cost functions are used at different stages of the annealing to improve the efficiency of the annealing since many solutions are searched. Running times were achieved comparable to heuristic techniques. However the rate of convergence to a global optimum (Nemhauser, 1988) is exponential. It was stated that new constraints could be added by changes to the cost functions.

3.5.3 Makespan Scheduling

A graph theory approach to the simultaneous scheduling and resource (modules and registers) minimization problem (Pfahler, 1987) was researched. A two dimensional placement of the data flow graph where makespan (or execution time), graph height (number of modules), and modified cutwidth measurement (estimated number of registers) were defined was used to represent the scheduling problem. The problem is that the cutwidth which can be solved easily includes all edges in the graph and we only need the edges representing the variable lifetimes. Thus we need only consider the longest outdegree arc of each node to represent the lifetime of the variable. This is why a heuristic was needed to solve the problem, since minimizing the lifetime defining edge (maximum length of all edges incident to a node) is NP-complete. A heuristic was used to solve this multiprocessor makespan scheduling problem.

3.5.4 Feasibility Models

The need for early area and delay prediction of different architectures for an input algorithm is very important. Some tools were developed to try to predict these performance values. The tools had to be very fast, and the current synthesis tools could not appropriately be used because significant amounts of time would be required to synthesize designs and subsequently calculate area and delays. Furthermore as a design exploration tool the synthesis would have to be done over a full range of the design space which would take too long. Thus feasibility models were created to help early prediction and to enable a better judgement of which area of the design space curve should be explored in detail.

Feasibility models for nonpipelined and pipelined architectural synthesis have been studied in (Jain, 1988) using simple mathematical equations for analysis before synthesis to narrow the design search space of interest. However these models only take into consideration the number of modules to be used in the architecture. The functional units, registers or interconnect were not considered in the mathematical equations.

3.6 TIMING CONSTRAINED SYNTHESIS

Timing constraints, as discussed in chapter 2, are very important for architectural synthesizers, even though few synthesizers (Nestor, 1986, Nestor, 1990) can handle these simultaneously with allocation subtasks. Not only are these important for supporting interfaces to external environments but they are also necessary for handling local application specific constraints within the synthesized architecture itself. For example timing constraints are required to model functional pipelining or possibly for multicycled operations.

The first synthesizer to consider timing constraints was Elf (Girczyc, 1985) where a timing constraint for a group of operations was specified. This constraint was generally a minimum or maximum execution time to be met.

More recently the Carnegie Mellon University synthesis effort has updated the CSTEP scheduler to incorporate minimum and maximum timing (Dutt, 1990) constraints. These constraints can be placed between any pair of operations in the algorithm. The list scheduler uses priority values for operations to decide if they must be placed in a certain control step. Timing constraints are checked and if a constraint is about to be violated by an operation not being placed in a control step then the priority value for this operation is modified to prevent the illegal assignment from being made.

Systems level partitioning research, APARTY, in (Lagnese, 1989) evaluates different partitions for the DAG but schedules the entire DAG without partitions. If the user requests two processes from the partitioner it will pass each partition separately to the scheduler but no responsibility for timing between the two processes is done and timing constraints are not used.

Research at Stanford University (Ku, 1989a, Ku, 1989b) has examined timing constraints for high level scheduling and logic synthesis. They identify a fixed timing constraint and a unknown unbounded timing constraint. It is assumed that module binding and hardware allocation has already been done, and an iterative algorithm for relative scheduling is presented. The feasibility of timing constraints is defined and an algorithm is also presented.

Other CAD areas which have identified timing constraints is logic or controller synthesis and design representation. Timing constraints and their effects on loops and conditional codes (Hayati, 1989) for a logic synthesis environment has been investigated. Asynchronous circuit synthesis in (Borriello, 1988) or (Meng, 1989) has also been researched but no datapath is synthesized. Design representation in (Dutt, 1990) has researched the use of charts to partition synchronous from asynchronous circuitry and perform partial binding of hardware. Other data representation such as the DDS in (Knapp, 1983) can be used to model both data and timing information. Finally in (Leiserson, 1970) interface timing constraints are modeled by using an external node called the host. A time t after or before the clock tick are used as constraints.

3.7 COST FUNCTIONS FOR DESIGN EVALUATION

The cost function is very important in architectural synthesis since it will influence the choice of the optimal architecture for a particular application. Unfortunately it is not clear what form this cost function should take. Ideally we want to minimize some area and delay cost function. The area cost can be estimated as some function of the number of functional units, registers, and busses (Devadas, 1989) . Assuming we have a model for estimating area (before placement and routing), we have to weigh this against the delay factor. For architectural synthesis the delay can be the number of control states to execute the algorithm or a more detailed value. The area cost and delay cost are two criteria. One now has to assign a weight to each and sum these to form an objective function. It is not clear how to weigh one over the other. Therefore research in multiple criteria optimization is relevant and very important for architectural synthesis.

Design evaluation with BUD (McFarland, 1987) showed that the area-delay curves vary a great deal when multiplexors and also layout and wiring are considered. The BUD algorithms used a cluster tree to provide a floorplan from which designs could be evaluated using a linear cost function. Their research in (McFarland, 1987) completed designs to layout to obtain accurate area-delay curves. Other research has incorporated floorplanning into design synthesis (Gebotys, 1989, Peng, 1987) Unfortunately placement and routing routines have not been modeled extensively to provide an area-delay model given a netlist or the allocated number of functional units, registers, and busses. of characterized hardware modules, however it is believed (Devadas, 1989, McFarland, 1987) that these models will be nonlinear.

In summary we have briefly discussed the different (locally optimal) approaches to state of the art architectural synthesis. The optimization of independent subtasks (of architectural synthesis) was shown to be limited for certain cases where the graph (obtained from the scheduled DAG) had a particular structure. It was also shown to be very difficult to extend this approach using graph theory for simultaneous solutions of more than one subtask. The previous integer programming approaches either were too large, and could not be solved, or were formulated to solve only a small part of architectural synthesis. Because of these complexities and the fact that architectural synthesis is most likely NP-hard, many researchers have turned to heuristics. In the next chapter we will discuss the recent successes in integer programming research. In particular this research involves the study of polyhedral characteristics and their use in the solution of large scale integer programming problems. Secondly we will show that unlike graph theoretical techniques even constraints with no apparent structure can often be solved using these techniques. Upon completing the next chapter the reader will be exposed to all the necessary background in architectural synthesis and integer programming necessary for the remainder of the text.

4.

INTRODUCTION TO INTEGER PROGRAMMING

General integer programming (IP) applications and solutions are briefly reviewed in this chapter. Section 4.1 outlines general formulation techniques for IP. Section 4.2 discusses state of the art solutions of general IP problems including classical enumerative and heuristic approaches (ie. simulated annealing). Recent successes in polyhedral approaches to solving partially structured IPs are outlined in section 4.3. Finally the definition and partial structure of the node packing problem (the focus of architectural synthesis) is given in section 4.4. (The notation for a graph is G=(V,E), where V is the set of vertices and E is the set of edges).

4.1 APPLICATIONS AND MODELS

Integer programming has an extremely large number of potential applications. Many VLSI design problems can be formulated as an IP problem and consequently there is a great deal of interest in this technique. Two important steps in integer programming are preprocessing and model formulation. Both the amount of preprocessing that can be done and the formulation of the model has a great impact on the final IP accuracy and solution efficiency. We will first look at one of the most simplest models, the assignment problem, that has many applications. A simple method for formulating constraints that can be represented as logical inferences is discussed next, followed by the definition of disjunctive constraints.

The assignment problem is one of the easiest models to formulate. The variables of the model are binary and each represents the mapping of i elements to j elements. For example figure 4.1a) illustrates a possible mapping choice, where the variables are the edges of the graph, $e_{i,j}$. If $e_{i,j}$ is 1, in the solution, then the assignment of i to j is optimal. Otherwise, if the value is 0, there is no assignment produced by the solution. Although we have used a bipartite graph [3] for illustration this type of assignment or matching is not restricted to these types of graphs alone.

A perfect matching problem is a set $M \subset E$ such that each node is incident to *exactly* one edge of M. The binary variables are:x_e=1 if $e \varepsilon M$ or x_e=0 if e is not a member of M. Thus we wish to solve the following optimization problem, where $\delta(u)$ is the set of edges incident to vertex u. Chapter 4 will further discuss this optimization problem in the context of polyhedral characteristics.

---

3 Bipartite graph is a graph with no odd cycle. It can always be partitioned into two groups X and Y (or i,j in figure 4.1).

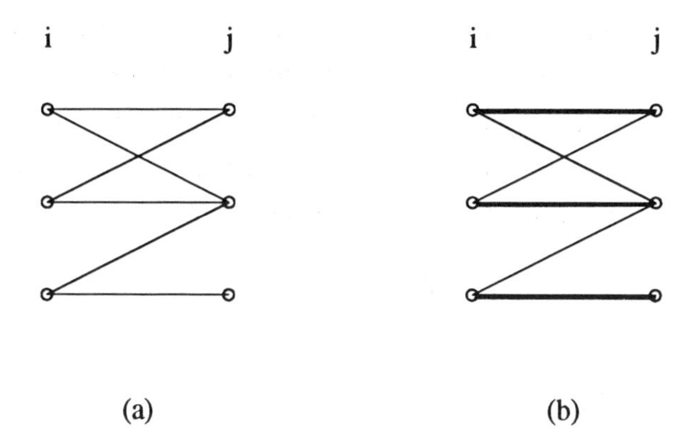

Figure 4.1. An assignment problem illustrated by a bipartite graph (G=(V,E)) with two partitions i and j. A solution, $M \subset E$ is shown in (b).

$$Max \ \ cx$$

$$\sum_{e \,\varepsilon\, \delta(u)} x_e = 1. \ \ \forall u \varepsilon V, \ \ x \varepsilon \{0,1\}.$$

A representation of logical inferences by mathematical linear inequalities has been examined by (Grossman, 1990) or (Ra, 1990) . For example the logical expression or inference $P_1 \Rightarrow P_2$ is equivalent to: 1) $\neg\, P_1 \vee P_2$ (Clocksin, 1984) and ; 2)$1\text{-}p_1\text{+}p_2 \geq 1$ or $p_1 - p_2 \leq 0$ (Ra, 1990) , where p_i are binary variables. For example if $p_1 = 1$, then for the inequality to be satisfied, p_2 must also be 1, which is the same as $P_1 \Rightarrow P_2$. Another example is $\neg\, y_1 \vee \neg\, y_2 \vee z$ which is equivalent to the mathematical inequality (4.1).

$$y_1+y_2-z \leq 1. \tag{4.1}$$

Integer variables can also be used to represent disjunctive constraints (Nemhauser, 1988) or model the activation or deactivation of a continuous variable. For example, y =1 $\Rightarrow L \leq x \leq U$ and y = 0 $\Rightarrow x$=0, can be modeled by the inequality (4.2). This represents a disjunctive constraint on x or a (de)activation of a continuous variable x by a binary variable y.

$$Ly \leq x \leq Uy \tag{4.2}$$

4.2 SOLUTION OF UNSTRUCTURED IPs

We will now look at a few general techniques for solving IPs with no apparent structure (see chapter 4.3 for more details on structure). These IPs are called unstructured IPs. The first step to solving an IP is to transform the IP into a relaxed LP and solve the LP. We transform an IP into a relaxed LP by removing the integrality constraints on the variables and allowing them to be solved as real positive numbers. For example we can replace $x_e \varepsilon \{0,1\}$ with $1 \geq x_e \geq 0$. If we obtain an all integral solution then we have found an optimal solution to our problem. Proof that the solution is globally optimal comes from the duality theory of LPs (Nemhauser, 1988) because we are solving the IP as an LP. In our LP solution if one or more variables are not integral then we have to look for other procedures to solve for the integral variables. This section will address this problem. We will assume that we are solving for binary variables (since any integer variable can be represented by a sum of binary variables).

We will first define some IP terms commonly used. There exists a *bounded polyhedron* for any rational bounded system of linear inequalities. Figure 4.2a) gives an example of a polyhedron defined by its constraints, $Ax \leq b$. We will call the convex hull of integer vectors an *integral polyhedron.* This is also illustrated in figure 4.2b), where the linear inequalities (now called *facets*) intersect at integer values (represented by the dots). These facets are of dimension one less than the dimension of the polyhedron. It was proved that for any bounded system of rational linear inequalities there exists an integral polyhedron, and in fact the facets are linear combinations of the inequalities defining the polyhedron. Unfortunately for most problems we do not know how to form these linear combinations or in other words we do not know what the facets look like. Furthermore even if we did there may be an exponential number of them. A final term to define is a *cut*. A cut is a valid linear inequality that cuts away fractional values from the existing linear programming fractional solution. For example in figure 4.2c) the dotted lines represent cuts.

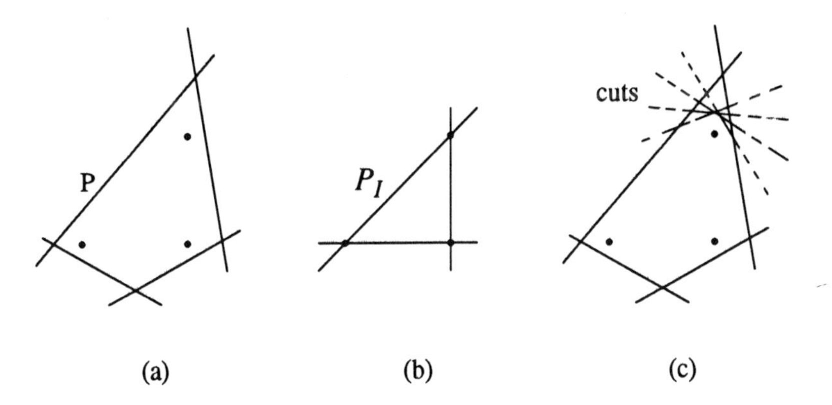

Figure 4.2. (a) illustrates a bounded polyhedron, (b) shows the corresponding integral polyhedron, and (c) identifies possible cuts, on the polyhedron of (a), as dotted lines.

General IPs may be difficult to solve (Nemhauser, 1988) due to 1) size of the formulation, 2) weakness of bounds, and 3) speed of the algorithm. For example in 1) the number of variables or constraints may be very large, in 2) the difference of the lower bound and optimal solution of a variables may be great, or in 3) the algorithm for solving the problem may be very slow. Recent success in solving IPs have shown that (in addition to preprocessing) by tightening constraints, or more effectively by using facets, (Nemhauser, 1988) one can dramatically improve the efficiency of solving IPs. We say that one constraint, $\delta x \leq \delta_0$, is tighter, dominates, or is stronger than the other constraint, $\zeta x \leq \zeta_0$, if $\{x \varepsilon R \mid \delta x \leq \delta_0\} \subset \{x \varepsilon R \mid \zeta x \leq \zeta_0\}$. One way to show this is to In other words let the polyhedron generated by the first set of constraints be P^1 and P^2 for the second set of constraints, then $P^1 \subset P^2$. One way to show this is to find a fractional point where $x \varepsilon P^1 \cap \overline{P^2}$, therefore $P^1 \neq P^2$, and (2) show that $P^1 \Rightarrow P^2$. The efficiency of solving the IP is improved due to the fact that tighter models have a smaller set of feasible solutions which must be searched. Branch and bound algorithms can be used to solve IPs in practical times if additionally the model has a small number of variables and tight bounds are known. The most well known general solution techniques for integer programming are the enumerative techniques such as branch and bound or heuristic variations. We will first review one of the oldest techniques for solving IPs, called Gomory's cutting planes algorithm.

Gomory's cutting planes is more interesting from a theoretical point of view than from a practical point of view. Generally Gomory was able to prove that after a finite number of cuts on any bounded polyhedron P, an integral solution can be obtained. He found a general method for obtaining these cuts using the simplex tableau of the LP solution. Unfortunately a very large number of cuts must be generated before an integral solution is found and few researchers use this technique on practical IPs because it takes too long.

The branch and bound method, or variation of it, may be used for a small number of variables (<200). However it is possible that even for small problems the solution may not converge due to the shape of the polyhedron. For the example shown in figure 4.3 a long narrow needle shaped polyhedron may require a long time to converge with branch and bound techniques. The intersection of dashed lines represents the integer values. The bound on the objective function is also very bad, for example the distance between X^* and X. The objective of the branch and bound technique is to create new LPs by bounding each variable towards integral values. The tree formed, by branching on a variable $x \geq \lceil X^* \rceil$ and $x \leq \lfloor X^* \rfloor$, is expanded only on nodes where the objective function is more optimum. From experience it has been found that an integral solution may be found quite early yet to finish the algorithm and therefore prove it is a global optimum takes a very large amount of time. Nevertheless it has been widely used for many small problems. Commercial software uses branch and bound techniques and can generally handle up to 200 integer variables (Brooke, 1988) .

There exist many heuristic techniques for solving IPs such as greedy algorithms, interchange heuristics, simulated annealing, and others (Nemhauser, 1988) . These techniques tradeoff optimality for efficiency. Tremendous success in solving many engineering problems with simulated annealing has been achieved, even though the convergence to a global optimum is exponential. Since combinatorial optimization problems have many local optima, some heuristic approaches, such as the greedy or interchange algorithm, are often run with random starting points. Simulated annealing is a different approach to avoiding local optima, by allowing the objective value to decrease only occasionally (for a minimization problem), to avoid getting stuck at a shallow local optimum and thus escaping towards another neighborhood with a smaller objective

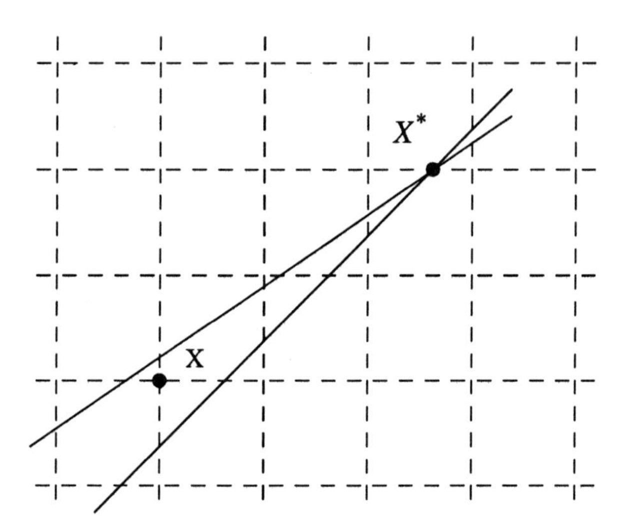

Figure 4.3. An example polyhedron that may take a long time to converge using a branch and bound techniques due to the needle shape of the polyhedron.

value.

A geometry of numbers approach (Cook, 1990) to solve particular IP's that cannot be solved using branch and bound has been researched. Generally IP's with not necessarily a large number of variables but those which exhibit a long needle-like polyhedron, as illustrated previously in figure 4.3, were solved using geometrical transformation.

Using a quadratic potential function projected on a ellipsoid the recent work of Karmarkar (Karmarkar, 1990) has shown that large sized integer problems known as the satisfiability problems can be solved. However if an objective function is required only a locally optimal solution is possible and there exists no guarantee of finding a solution. Thus this approach seems to be directed towards a problem characterized by a small number of integral optimal solutions.

4.3 POLYHEDRAL APPROACHES TO SOLVING IPs

In general solving an IP problem is NP hard (Garey, 1979) . However, analogous to special graphs in graph theory, there exist special techniques for solving some IPs. Thus all IPs are not equivalent in difficulty in all respects. For example to solve a node packing IP problem on a graph which is claw-free (ie. ∃ no $K_{1,3}$ [1]) requires only polynomial time, using Minty's (Minty, 1980) algorithm. This is analogous to the graph theory approaches where polynomial algorithms are known to exist if the graph at hand is of a particular structure (ie. interval graph for polynomial time algorithms that perform node coloring (Golumbic, 1980)). We say that these IPs have *structure*. Additionally IPs where some constraint has this property are said to have *some structure*. In IP we can often obtain good bounds on a particular problem and often solve for integer variables using this structure, even when no known graph theoretical algorithms, heuristics or formulations may exist. But how can we find this structure? We can often do this through proper model formulation.

The research focus over the past 25 years in IP has been to study polyhedra characteristics of a problem and thus define structure which may help in its solution. This was motivated by the desire to obtain tight formulations of the problems rather than adhoc models, since IPs have exhibited extremely erratic performance. A systematic way to obtain these formulations is to analyze facets. Unfortunately there exists no formal method for obtaining facets of a given IP and even if we could find a

---

1 $K_{x,y}$ is a complete bipartite graph with partition x,y.

method to generate all facets, most likely we couldn't solve the LP because there may be an extremely large number of them (possibly exponential). Balas and Padberg (Hammer, 1979) have argued that its very useful to find facets or approximation of facets because only a few define optimal points. Also it is known that if one used a branch and bound technique after extracting some facets, the algorithm would generate fewer live nodes (Padberg, 1979) and terminate faster. This is mainly due to the better bound obtained from the use of facets. Thus by mapping a problem or subsets of a problem into a well studied class of problems, such as node packing, whose facets are partially characterized one may be able to improve the bounds of the problem and solve for integer variables more efficiently.

Recent research has proven how important facets are. The tremendous success of the use of facetial characteristics is demonstrated with the traveling salesman problem (Lawler, 1985) and large sparse unstructured IPs solved by using facets of subproblems in (Crowder, 1983) etal. Further research (Lawler, 1985, Crowder, 1983) has also shown how it is highly advantageous to add facets to the LP until no new ones can be found even before you start to branch and bound.

State of the art solutions of unstructured IP have been researched by (Crowder, 1983) using a combination of preprocessing, cutting planes (using knapsack facets of underlying polytopes), and branch and bound techniques to solve sparse 0-1 unstructured IPs of over 2000 variables in reasonable computation times (less than 1 cpu hour). The cutting planes which were facets of the underlying polytope (knapsack inequalities) were extremely useful and successful for exact solution of their class of problems. Their system was completely automatic, and represents state of the art for solving unstructured IPs. When a cut cannot be found a variable is selected to branch on. The definition and characterization of knapsack inequalities is given in section 4.5.

In 1980, Grotschel (Grotschel, 1980) demonstrated optimal solution of (over 7,000 integer variable) TSPs in 30 cpu sec to 2 cpu min to show the usefulness of the theoretical research in polyhedral characteristics. In all cases the problems could not be solved using existing branch and bound techniques, thus demonstrating the importance of polyhedral combinatorics in solving large scale optimization problems. In 1980 Padberg (Padberg, 1980) solved for 50,000 integer variables of the TSP problem completely automatic to within 0.25% optimality in 30 minutes using automatically generated facets. Unfortunately the number of applications which can be modeled as a traveling salesman problem is not proportional to the large amount of research that this problem has generated. Conversely there are other problems, such as finding the maximum weighted directed cycle in a graph that have a large number of applications, but generated little research. This is also partially true for the node packing problem in a smaller sense as we shall see in section 4.4.

4.4 THE NODE PACKING PROBLEM

There exists a great deal of interest in the node packing problem because of a) the large number of practical applications and b) the stronger structural properties than the general integer programming problem (Padberg, 1973) . The node packing problem has also been called vertex packing and the stable set problem. It is also related to other problems in optimization such as the set covering, set packing, anticliques, independent sets, and node covering, (Padberg, 1973, Nemhauser, 1974) which we will not cover in this text. We will first illustrate the relationship between integer programming, graph theory, and node packing, using a simple completely structured problem (that of maximum matching). Secondly we will formally define the problem and then proceed to define the known facets of this problem.

Integer programming and graph theory have many areas of research which overlap. For example figure 4.4(a) illustrates a perfect matching problem. Each edge must be assigned a 0 or 1 value to maximize the sum of all edges with the restriction that each vertex is incident to at most one edge with a value 1. We can alternatively use the Hungarian Method or Kuhn Munkres (Bondy, 1976) algorithm to solve for a maximum matching in polynomial time. Alternatively one can solve an IP where constraints correspond to integral facets. In the later method we can solve the IP as an LP and be guaranteed to always obtain a solution with integer variables. The second constraint given below can be automatically generated as needed for a particular problem by at most 2n-1 min cut problems on the graph. In other words instead of generating this constraint for all odd sets of vertices we can solve the LP and automatically generate facets to cut away the fractional values and solve for integer variables using the relaxed LP. The complete model for weighted perfect matching is given below, where $\delta(S)$ is the set of edges, where each edge has one vertex in S and the other vertex in $\bar{S}$.

$$Max \;\; cx$$

$$\sum_{e \varepsilon \delta(u)} x_e = 1, \;\; \forall u \varepsilon V.$$

$$\sum_{e \varepsilon \delta(S)} x_e \geq 1, \;\; \forall S \subset V, |S| odd.$$

$$x_e \geq 0, \;\; \forall e.$$

The vertex representation of this problem (which we will define later as node packing) is shown in figure 4.4(b) where each edge is now a vertex (variable) and edges of this new graph represent adjacent vertices of the matching graph in (a). The graph in figure 4.4(b) is a line graph obtained from (a), and it is known that the solution of this problem (node packing) on a line graph (Nemhauser, 1988) can be solved in polynomial time. This example briefly illustrates the relationship between graph theory and integer programming (and node packing). However this relationship does not hold true for all cases. There exist some problems for which known polynomial algorithms exist (ie. it is well solved) however the associated polyhedron is nontrivial. An example of this is to find the edge in a node weighted graph whose sum of weights of its two incident vertices is maximum (Hammer, 1979) .

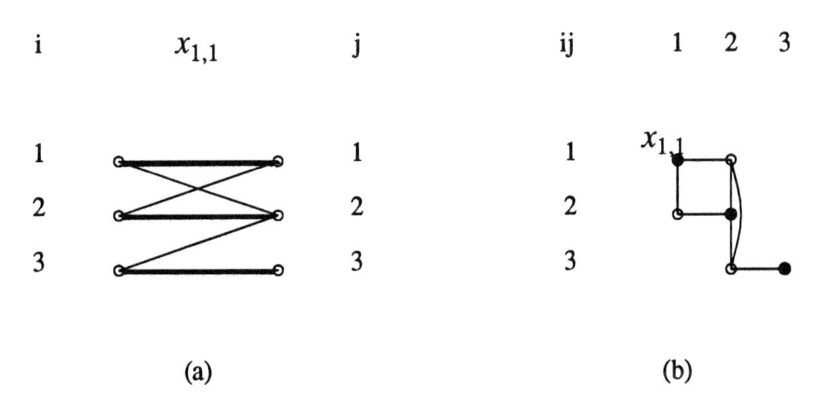

Figure 4.4. The matching problem represented by edge variables in (a) and vertex variables in (b). In the former case one assigns 0 or 1 to edges and in the later case one must solve a node packing problem, by assigning 0 or 1 to vertices.

Like the traveling salesman problem characteristics of the integral facets are partially known (Nemhauser, 1988) for the node packing problem. This problem is more formally stated below in two forms. One form is the graph theoretical view and the second is the mathematical linear system of equations view.

1. In graph theoretical form: Given a graph G = (V,E), maximize cx, such that

$$x_u + x_v \leq 1, \ \forall (u,v) \, \varepsilon \, E$$

$$x_u \geq 0, \ \forall u \, \varepsilon \, V.$$

2. In linear systems of equations form:

$$\max cx$$

$$Ax \leq e$$

$$1 \geq x_j \geq 0, \forall j \, \varepsilon \, N = \{1,...,n\}$$

where A is a mXn node edge (0,1) incidence matrix, c an arbitrary n-vector, and $e^T = (1,...1)$ is an m-vector (Padberg, 1973)

If all variable solutions are integral then a globally optimum solution to the problem has been found and we are done. A property unique to the node packing problem is that if not all variables solutions are integral, the variables that are integral remain integral (Nemhauser, 1988) in the optimum solution. Therefore the problem can be decomposed into a smaller problem to solve. However it is also known that this node packing formulation with node edge incidence constraints, generates very poor bounds (Padberg, 1979) . Furthermore studies which attempt to use this property to solve the problem have found that in most cases few integer variables are attained (Grimmett, 1985) . We will discuss the node packing problem using the graph theoretical formulation.

Finding all integral facets for a particular node packing problem is NP-complete. This problem is known as the *stable set polytope* (SSP) problem, using graph theoretical terminology. Nevertheless only integral facets over the region of the minimum objective function are required to obtain integral solutions. We will now define some of these facets.

Known integral facets for the node packing problem are given in (4.3) and (4.4).

$$\sum_{u \varepsilon K} x_u \leq 1, \textit{for all K cliques} . \qquad (4.3)$$

A clique (or maximal complete subgraph) is a subset of nodes K for which there exists an edge in the graph for every pair of nodes in K.

$$\sum_{u \varepsilon C} x_u \leq (|C|-1)/2, \textit{for all C odd cycles without chords}. \qquad (4.4)$$

An odd cycle is an odd number of nodes in the graph, which form a subset, C, of the graph, such that the edges form a cycle. Without chords means that no 2 nodes of C can share an edge that doesn't belong to the cycle. Normally we use the term odd cycle for graphs with 5 or more nodes (connected in an odd cycle), and the term clique is used to describe the three node case (which also forms an odd cycle). In addition some odd cycles may be "lifted". The term lifting refers to placing a neighbour node (a node connected to several nodes of the odd cycle) into the inequality with a positive coefficient c_i where $c_i \geq 1$. This property makes node packing a more difficult problem than the matching problem where all coefficients of the integral polyhedron are 0 or 1. An example of lifting an odd cycle, from $a+b+c+d+e \leq 2$ to $a+b+c+d+e+2f \leq 2$, is given in figure 4.5(c). Whenever node f is equal to one then all other nodes (a through e) must be zero therefore the odd cycle in 4.5(c) can be lifted by adding node f to the inequality with coefficient 2 (equal to the right hand side value). If edge (f,d) were removed from the graph of figure 4.5(c) then the coefficient of f in the inequality would be 1.

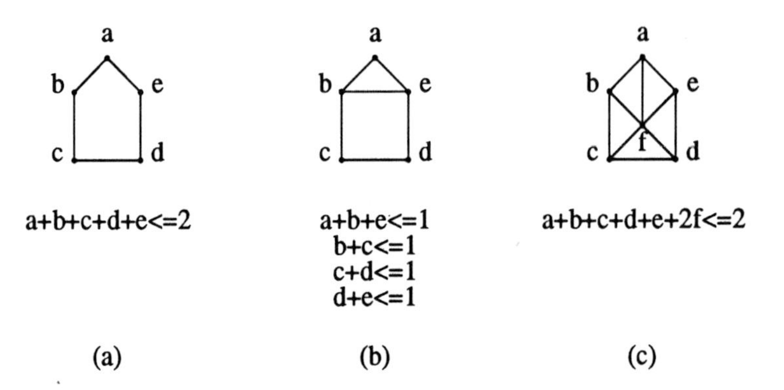

Figure 4.5. Node packing on the three graphs: an odd cycle facet in (a). An odd cycle with a chord which reduces to 3 edge inequalities and one clique facet in (b). (c) shows a lifted odd cycle facet.

4.5 THE KNAPSACK PROBLEM

In many IPs some constraints may fall into the category of knapsack inequalities. By generating known facets of this underlying (knapsack) polytope, one can often tighten the larger polytope represented by all inequalities. This has been very successful as demonstrated by the award winning paper in (Crowder, 1983) . The definition and facet characterization of the knapsack problem will be given in this section.

Consider the following polytope,

$$P=\{x \mid \sum_{j \varepsilon N} a_j x_j \leq b, 0 \leq x_j \leq 1\}, 0 \leq a_j \leq b, \forall j \varepsilon N. \tag{4.5}$$

Minimization (or maximization) of some cost function subject to the inequality (4.5) is known as the knapsack problem. We wish to find P_I, the integral polyhedron for this problem. If we find the integral polyhedron then we can solve for the binary x variables for any cost function.

We will now introduce some notation. Let $\bar{x}$ be an integer vector in P. Then we can represent this fact by saying that the set $S=\{j \mid \bar{x}_j=1\}$ is independent. In other words we say that a x vector is independent if it satisfies the inequality in (4.5). Now let C be a minimal dependent set. In other words we say that a x vector is dependent if it does not satisfy the inequality (4.5). The dependent set C refers to the set of subscripts j of the x vector such that x_j =1. The dependent set C is minimal if and only if C \ {i} is independent ∀ i ε C. In other words C is minimal if all of its subsets are independent. The following inequality (4.6) is valid for P_I.

$$\sum_{j\varepsilon C} x_j \leq |C|-1. \tag{4.6}$$

Now let us assume that in (4.5) $a_1 \geq a_2 \geq \cdots \geq a_n$. Given C, we can define E(C)= C $\cup \{k : a_k \geq a_j, \forall j \varepsilon C\}$.

Now the following inequality (4.7) is valid for P_I and it is tighter than (4.6).

$$\sum_{j\varepsilon E(C)} x_j \leq |C|-1. \tag{4.7}$$

The inequality (4.7) is a facet of P_I if and only if at least one of the four conditions given below are true.

1. C = N
2. E(C) = N and (i) $C \backslash \{j_1, j_2\} \cup \{1\}$ is independent.
3. C = E(C) and (ii) $C \backslash \{j_1\} \cup \{p\}$, $p=min\{j \mid j \varepsilon N \backslash E(C)\}$ is independent.

4. $C \subset E(C) \subset N$ and (i) and (ii).

Unfortunately if the a_j's are 0,1,-1 these facets are not of much use. Generally if the a_j's do vary in magnitude the facets are very useful. We will show how important these facets are for the application of architectural synthesis in chapter 10.

This chapter has presented a brief look at research in integer programming. In particular polyhedral characteristics and their use for solving structured and unstructured problems was emphasized. Due to the erratic behavior of integer programming problems, the use of polyhedral characteristics of a problem has a significant impact on solving the IP efficiently. Sufficient background material has now been presented to introduce the OASIC (for optimal architectural synthesis with interface constraints) methodology in the next chapter. Node packing facets, tightened constraints, and the use of knapsack inequalities which were introduced in this chapter are used in chapters 6 through 9 for the OASIC model.

PART III : OPTIMAL ARCHITECTURAL SYNTHESIS WITH INTERFACES

5.

A METHODOLOGY FOR ARCHITECTURAL SYNTHESIS

In this chapter we introduce the requirements for a high level synthesis tool and outline exactly what constructs OASIC (Optimal Architectural Synthesis with Interface Constraints) the high level synthesis tool, to be defined in chapter 6 and 7, can support. The high level systems design methodology and specific OASIC methodology are defined below. In summary we discuss the impact of the OASIC tool on industrial CAD needs.

5.1 REQUIREMENTS FOR HIGH LEVEL SYNTHESIS TOOLS

Architectural synthesis is an important part of the VLSI design cycle. The objective of synthesizers is to transform an input algorithm into a hardware architecture that satisfies a set of constraints and minimizes or maximizes a given cost function. Synthesizers must produce globally optimal architectures and execute quickly in order to provide early exploration of design tradeoffs. In addition synthesizers should be able to optimize linear or piecewise linear cost functions (for modeling area and delay), incorporate complex constraints (which may arise during design), interface to other hybrid processes (analog or asynchronous), and interface to tools for test incorporation. An architectural synthesizer's primary responsibility is to aid high level design exploration which includes systems level design (more than one chip), where a number of different analog/asynchronous or digital paradigms exist. Synthesizers must handle complex timing constraints for interfacing to 1) analog signal processing modules, 2) asynchronous modules (or data dependent operations), or 3) a different clocked domain of synchronous digital modules. Furthermore the DA synthesis tool should also support functional pipelining.

The OASIC synthesizer, which will be described in chapters 6 and 7, performs simultaneous scheduling and allocation of functional units, registers, and interconnect. The following features are supported:

- Timing constraints including minimum, maximum or fixed timing constraints.
- Behavioral interface to unknown bounded and unbounded timing constraints, which may represent i) analog signal processing units, ii) other asynchronous processes, or iii) data dependent operations.
- Piecewise linear area-delay cost functions
- Random topologies
- In addition to allocation, simultaneous selection of types of functional units.
- Pipelined, multicycle or single cycle functional units.

- Conditional code, implemented by sharing hardware resources, and loops.
- Globally optimal synthesized architectural solutions with respect to cost function.
- Functional pipelining for a fixed latency

5.2 HIGH LEVEL METHODOLOGY

A proposed formal methodology for high level systems design is shown in figure 5.1. The input to the methodology is a high level behavioral description, of the form described in chapter 2 or a mixed level (behavioral and structural) description of the system to be synthesized. The final result of applying the methodology is one or more chips with a mixed analog or digital implementation. The high level partitioning may be performed by a partitioning tool or possibly by hand to determine which design should be implemented in analog or digital and the later using synchronous or asynchronous logic. After the behavioral partitioning, the interface, as defined in chapter 2, between the digital synchronous and other analog or digital processes is defined. The mixed interface and behavioral specification is then input into the OASIC high level synthesis tool which maps the software into hardware. An optimal (and correct) schedule and allocation of hardware resources, forming an architecture, is synthesized by OASIC.

The stage after OASIC performs a second optimization which binds operations to hardware to minimize the number of bus drivers or bus connections. This binding problem is addressed in chapter 12 of part IV in the context of test incorporation. The future extension of OASIC involves the optimization of the binding phase simultaneously with test incorporation. The OASIC stage is very important and must be performed before the binding stage. OASIC minimizes the larger components of the design, ie. functional units, busses, and registers. Also since OASIC can solve for globally optimal architectures and schedules we believe that one can obtain solutions very close to the global optimum of the simultaneous scheduling, allocation, and binding problem, with respect to minimizing an area delay cost function. The later problem has never been formulated or solved. Note that the model formulation of OASIC can be extended for solving this and other similar problems as discussed in the future research part V. In fact OASIC is the first methodology to mathematically formulate and to solve simultaneous scheduling and allocation of all possible hardware resources to global optimums. In a systems design methodology this provides early design exploration. Once a designer decides upon which area-delay cost best suits her/his design application using OASIC, she/he can justify using more time to solve the binding problem and therefore obtain an optimal architectural design solution. In the next section we will look at a more detailed view of OASIC.

5.3 OASIC METHODOLOGY

In this section we will discuss the OASIC methodology in detail. Figure 5.2 illustrates the flow chart of OASIC. We assume that a DAG describing the behavior and interface constraints is input along with an area delay cost function. The OASIC methodology avoids early binding or making early design decisions which may be poor. An integer programming model is used to specify the function of an architectural synthesizer. Two models will be presented in chapters 6 and 7 and are illustrated in figure 5.2 as $x_{i,j,k}$ and $x_{j,k}$ respectively. The OASIC methodology can be divided into preprocessing stages, and optimization stages. The preprocessing stages provide the application specific data for the IP model. As soon as possible and as late as possible schedules are obtained first. This is a well known problem whose solution was discussed in chapter 3.3.1. A set of functional units can be selected by

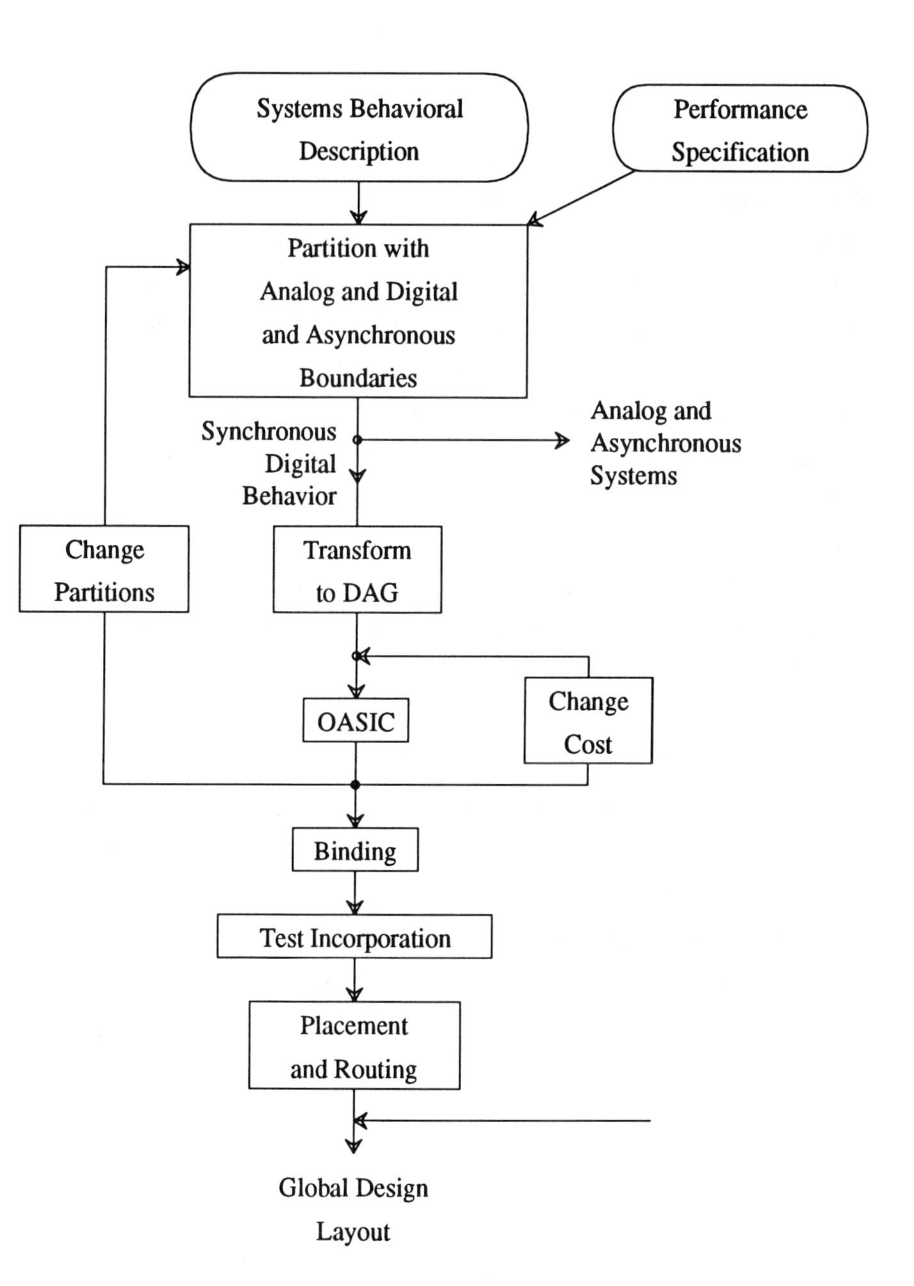

Figure 5.1. High Level Systems Design Methodology.

operation grouping or by the user. Other model specific preprocessing is discussed in section 5.5. The optimization phase can provide an early prediction or an optimized schedule and hardware allocation. The early prediction phase is used for early design exploration to study the area-delay characteristics of the particular input algorithm. This phase is important since the cost function is an estimate of the area-delay parameters. By varying the cost parameters one can explore optimal architectures. After the designer has decided on which architectures they are interested in, they can proceed to the solution phase of OASIC to obtain the complete schedule and allocation of hardware. The OASIC methodology is designed to avoid large amounts of feedback by using area-delay cost functions, solving for globally optimal solutions, and supporting direct interface to external or analog/asynchronous operations which may have complex timing constraints. These optimized early decisions are believed to have a significant impact on the final VLSI implementation thus decreasing the need for feedback.

5.4 AN INTRODUCTION TO OASIC

In chapter 4 we provided the definition and previous research for different subtasks involved in the architectural synthesis problem. Now we will identify the two major architectural synthesis problems for which we present a model in chapters 6 and 7. An exact definition for these problems in the context of architectural synthesis is given below.

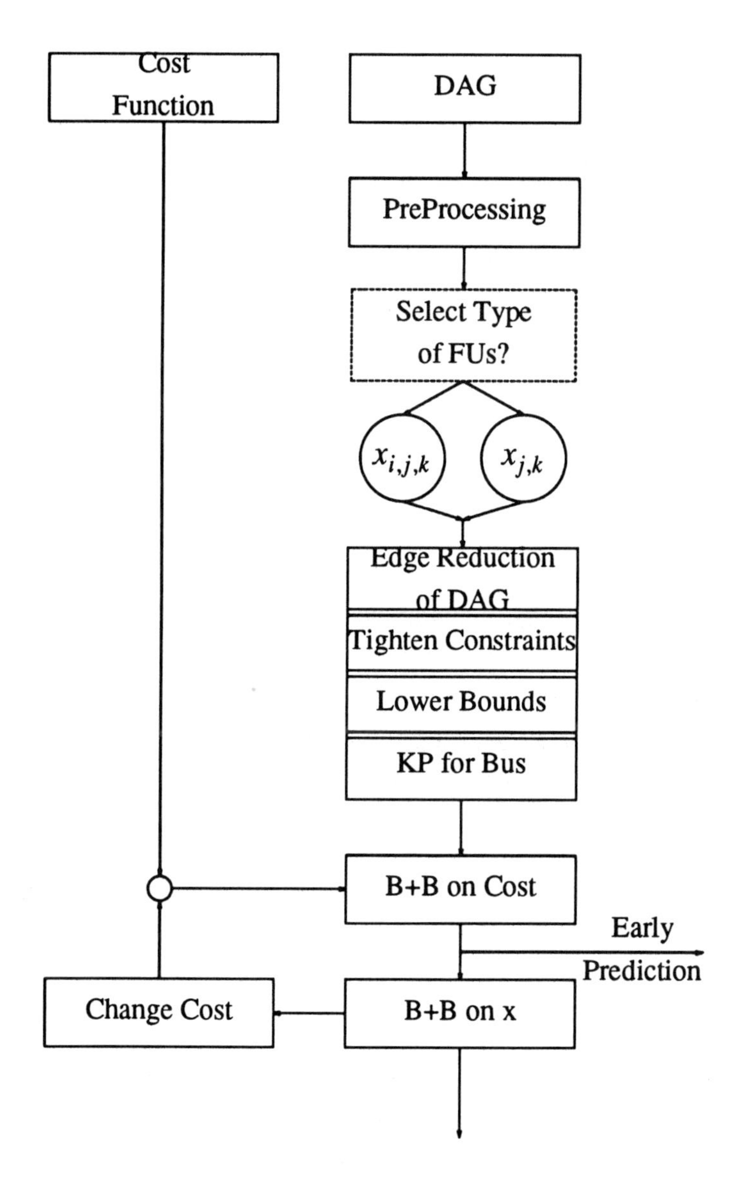

Figure 5.2. OASIC methodology for optimal architectural synthesis.

PROBLEM 1

Produce a schedule, by mapping each code operation to a time (maintaining the partial order among operations), and map each operation to a functional unit. Simultaneously select and allocate functional units, and schedule operations to minimize an area cost function.

The structured model in chapter 6 solves problem 1 (see figure 5.2 $x_{i,j,k}$). This problem has been called *simultaneous scheduling, and selection and allocation of functional units.*

PROBLEM 2

Produce a schedule, by mapping each code operation to a time (maintaining the partial order among operations) that minimizes a piecewise linear (area and delay) cost function of the number of functional units, registers, busses, and execution time (the total number of csteps required to execute the algorithm on the final architecture).

The area-delay model in chapter 7 solves problem 2 (see figure 5.2 $x_{j,k}$). This problem has been called *simultaneous scheduling, and allocation of functional units, busses, and registers.* Unlike problem 1 this model does not select a type of functional unit.

As discussed in chapter 3 the basis of both problems is precedence constraint scheduling. Our submodel for solving this subproblem is new and we will prove its advantages over previous research in the following chapters. Advances in computers providing faster computations and larger memories for mathematical software has also had a great impact on this modeling area.

The following algorithmic and complex constraints are also supported by OASIC's two models. The constraint formulations are presented in chapters 8 and 9.

> Additional Features to Support for Problems 1 and 2.
>
> The following features are to be supported: (1) Interface to analog and asynchronous processes, (2) Minimum and maximum timing constraints, (3) Conditional Code, (4) Functional Pipelining, (5) Make use of regularity and hierarchy, and (6) flexible piecewise linear cost functions.

Many subtasks of architectural synthesis, such as minimizing registers in the presence of general conditional code, can not be solved to global optima using previous algorithms. Thus not only is the larger problem of simultaneous scheduling and allocation being solved for the first time but many of its subproblems can now for the first time be solved optimally by OASIC. Furthermore as we shall demonstrate it is easy to incorporate these above features (shown in the box) into our model, whereas it may be difficult to make modifications to heuristics of previous synthesizers. We will show in chapter 10 that it is feasible to solve to global optima including these features in our model.

5.5 OASIC TERMINOLOGY, ASSUMPTIONS, AND PREPROCESSING

The terminology used and assumptions made for this mathematical model will be described in this section.

5.5.1 Terminology

1. k = code operation. A partial order (or precedence constraint) between k_1 and k_2 is represented by $k_1 <\bullet k_2$ (Garey, 1979) , or in other words k_1 must execute before k_2. Let K represent the total number of code operations in the input algorithm. Let A represent the number of arcs in the DAG.
2. j = a time, or a control step *cstep*, j=1,2,...,J^{UB}, J^{UB} = the upper bound on number of csteps given by the user. Te = total optimized number of csteps required to execute the DAG on the architecture.
3. i = a functional unit, i=1,2,...,$\sum_t I_t{}^{UB}$, I_t = the number of functional units of type t. $I_t{}^{UB}$ is the upper bound on the number of functional units of type t. For example t=1 (for adders) or t=2 (for alus). We will also use $k \varepsilon t$ to show that k can be mapped to functional unit type t.
4. m = a register, m=1,2,...,R^{UB}, R = the number of registers.
5. l= a bus, l=1,2,...,B^{UB}, B = the number of busses.
6. $x_{i,j,k}$ =1, represents the assignment of
   code operation k to functional unit i, at cstep j.
7. $j \varepsilon R(k)$ means that j is lower bounded by the *asap* scheduling time and upper bounded by the *alap* scheduling time for code operation k or $j \varepsilon \{ j_{asap(k)}, (j_{asap(k)}+1), \cdots, j_{alap(k)} \}$.
8. time(k_1, k_2) $\leq$ *or* $\geq$ *or* = T, denotes the (maximum or minimum or fixed) timing constraint between two operations, or subsets of operations. In other words the number of csteps between k_1 and k_2 is $\leq, \geq, =$ T.

9. $i_z \varepsilon Op(C_z, L_z)$, refers to the functional unit characteristics where C_z is the execution time (number of csteps from when the input data is ready to the time when the output data is available) , and L_z is the latency time (minimum time between successive data input values being accepted by the functional unit). We will use the following notation (which assumes a one to one mapping of operations to types of functional units), because it is easier to write the constraints of the model, where $k \varepsilon t$, the notation will be $k_z \varepsilon Op(C_z, L_z)$ as defined above.

10. In(t)= the number of inputs for functional unit type t, In(t) = 0,1, or 2. Out(t) = the number of outputs for functional unit type t, Out(t) =1 normally. Similarly for a one to one mapping we can use the notation In(k), where In(k) = In(t) | $k \varepsilon t$.

5.5.2 Assumptions

The current assumptions made in our model are listed below. Some of the extensions discussed in chapter 11 deal with removing these restrictions.

- Functional units that are chained must be specified by the user.
- Same bit width for bus allocation
- global and local code transformations (ie. loop unrolling) and partitioning of the DAG or input algorithm (into more than one chip) could be performed by other tools before OASIC.
- Global data broadcasts are specified in the DAG by the user (see chapters 6 and 7).

5.5.3 Preprocessing

Preprocessing that includes high level code transforms (such as those produced by optimizing compilers), conditional code extraction of blocks or branches, and regularity extraction are performed at a level higher than OASIC. This is consistent with current higher level tools such as SAW (Walker, 1987) ,that are user interactive. Architectural synthesizers are usually embedded underneath these tools. The only necessary preprocessing that must be done before input to OASIC is easily automated. This includes asap and alap scheduling and was discussed in chapter 3. After preprocessing the user must input their choice of types of functional units for allocation. Operation grouping is defined by the user before allocation to prevent illegal functional units from being allocated. An illegal functional unit is a group of operations which cannot all be performed by a single functional unit in the library. Our objective is then to identify legal code operation groups which may be executed by a hardware unit from an existing library.

Upper bounds on variables of the objective function are not required, however they can improve performance if specified (by the user) by decreasing the size of the search space. For example the area may not exceed a certain value or the number of clock periods required to execute one pass of the DAG (or to execute a conditional path) may not exceed a given number of cycles.

In this chapter we have described the OASIC architectural synthesis methodology in the context of a high level methodology for mixed technology systems design. In the next two chapters we will present the basic OASIC model. In chapter 6 selection of the type of functional units is addressed, and facets of the node packing polytope are extracted. Chapter 7 presents a complete model obtained by trading off structure for a reduction in the number of variables. Facet extraction of subpolytopes and tightening of unstructured constraints are used to form the final model. The use of the OASIC model for supporting general algorithmic constructions such as functional pipelining and for supporting behavioral interfaces are covered in chapters 8 and 9.

6.

SIMULTANEOUS SCHEDULING, AND SELECTION AND ALLOCATION OF FUNCTIONAL UNITS

A model for solving problem 1 of chapter 5 (section 5.4), simultaneous scheduling and selection and allocation of functional units, will be presented in this chapter. In general the problem is modeled as an assignment problem, where the variables represent a placement of code operations in two-dimensional space. The two-dimensional space is defined by time (in terms of control steps) and area (in terms of functional units).

The OASIC assignment model is not unlike that used in the simulated annealing algorithm (Devadas, 1989) where our $x_{i,j,k}$ variable represents the placement of operations on a two-dimensional grid. Thus the explosive nature of the problem is the same. However, in the simulated annealing approach, they do not allocate interconnections but use a heuristic measure (the number of parallel data transfers counting only the number of distinct sources for broadcasts data transfers). In addition the calculation of number of registers is included only in the cost function, and a mathematical formulation is not given. In addition the cost function is used to eliminate illegal solutions that their approach will encounter, whereas we use preprocessing to eliminate illegal designs before the design search begins.

In Pfahlers graph representation (Pfahler, 1987) interconnection is also not considered and proper calculation of registers cannot be performed. Finally the OASIC model was developed because it was very simple to incorporate interconnect, in addition to complex timing constraints, to be discussed in chapter 9. Incorporation of interconnect allocation has not been performed by any other simultaneous approaches. Only heuristics to estimate interconnect has been examined by previous methods (Devadas, 1989, Paulin, 1989) .

6.1 THE FORMAL MODEL

We will present the initial IP model formulated for solving problem 1. We will show how one can translate this model into the node packing problem. Using the node packing graphs for simple DAGs, extracted facets were generalized and used to form the IP model for OASIC. The IP model uses variables $x_{i,j,k}$ to represent the two dimensional placement of code operations and consists of three types of general constraints discussed below.

Equation (6.1), called the operation assignment constraint, ensures that each code operation will be assigned to one control step, functional unit, and register.

$$\sum_{i} \sum_{j \varepsilon R(k)} x_{i,j,k} = 1 \, , \forall k. \tag{6.1}$$

Inequality (6.2), called the functional unit constraint, prevents more than one code operation from being assigned to the same functional unit at the same control step.

$$\sum_{\substack{k \\ k \, \varepsilon \, Op(C,L)}} \; \sum_{\substack{j_1=j \\ j_1 \, \varepsilon \, R(k)}}^{j_1=j+L-1} x_{i,j_1,k} \leq 1 \, , \forall i \, , j. \tag{6.2}$$

Inequality (6.3), called the precedence constraint, prevents an operation, k_1 from being scheduled after operation k_2 whenever there is a partial order between these operations such that $k_1 <\bullet\, k_2$.

$$\sum_{i} (x_{i,j_1,k_1} + x_{i,j_2,k_2}) \leq 1, \tag{6.3}$$

$$\forall k_1 <\bullet\, k_2 \, , j_2 \leq j_1 + (C_1 - 1) \, , j_1 \varepsilon R(k_1) \, , j_2 \varepsilon R(k_2) \, .$$

$$x_{i,j,k} = 0 \; or \; 1. \tag{6.4}$$

When we relax (6.4) to $0 \leq x \leq 1$, the system of linear inequalities becomes $Ax \leq b$, where A is a (0,1) matrix (or all entries in A are 0 or 1) and b is a unit vector. This problem, known as set packing (Nemhauser, 1988) , can be transformed into a graph, G=(V,E) (called the node packing graph), from which we can extract some integral facets of the node packing problem. First we map each variable, $x_{i,j,k}$, into a vertex, $u \varepsilon V$. Edges of the graph, E, are defined by the following procedure. For each row of the matrix A (representing a constraint), the variables (or columns) with a '1' entry define a complete subgraph [2] (see chapter 4) in G (Nemhauser, 1988) The graph G represents the node (or

2 A complete subgraph is a subgraph $K \subset G, K=(V^1,E^1)$, such that $\forall$ $u,v \varepsilon V^1$, $(u,v) \varepsilon E^1$.

vertex) packing problem, which can be transformed from any system of linear inequalities with a (0,1) constraint matrix (A) and a unit vector (b) on the right hand side.

Known integral facets for the node packing problem are (as outlined in chapter 4) cliques and odd chordless cycles.

$$\sum_{u \varepsilon K} x_u \leq 1, \forall K \ cliques. \qquad \text{(i)}$$

$$\sum_{u \varepsilon C} x_u \leq (|C|-1)/2, \forall C \ chordless \ odd \ cycles, |C| \geq 5. \qquad \text{(ii)}$$

and others in (Nemhauser, 1988) which involve a lifting procedure for odd cycles. For the architectural synthesis problem, clique (integral) facets were extracted from node packing graphs representing some small input algorithm examples. The facets were then generalized and placed in a relaxed LP model. The LP model was applied to some benchmark architectural synthesis examples and optimized for different cost functions.

The one-dimensional model ($x_{j,k}$) is used to illustrate the facet extraction and generalization. This model can be visualized as the placement of code operations into control steps. For example consider two code operations with a partial order between them, $\{a,b \,|\, a <\bullet b,\ a,b \varepsilon Op(1,1)\}$, and an upper bound of 5 control steps. The node packing graph is shown in figure 6.1. The edges are formed in the graph using inequality (6.3), or in other words all illegally scheduled combinations of a and b form edges in the graph. Some clique facets of this graph are shown in bold in figure 6.1a) and b). We can generalize this clique facet to (6.5), which we will call the precedence or partial order constraint.

$$\sum_{\substack{j_1 \le j+C_2-1 \\ j_1 \varepsilon R(k_1)}} x_{j_1,k_1} + \sum_{\substack{j \le j_2 \\ j_2 \varepsilon R(k_2)}} x_{j_2,k_2} \le 1, \tag{6.5}$$

$$\forall k_2 <\!\bullet\, k_1 \,,\, j \,\varepsilon\, R(k_1) \cap R(k_2) \,, k_2 \,\varepsilon\, Op(C_2, L_2).$$

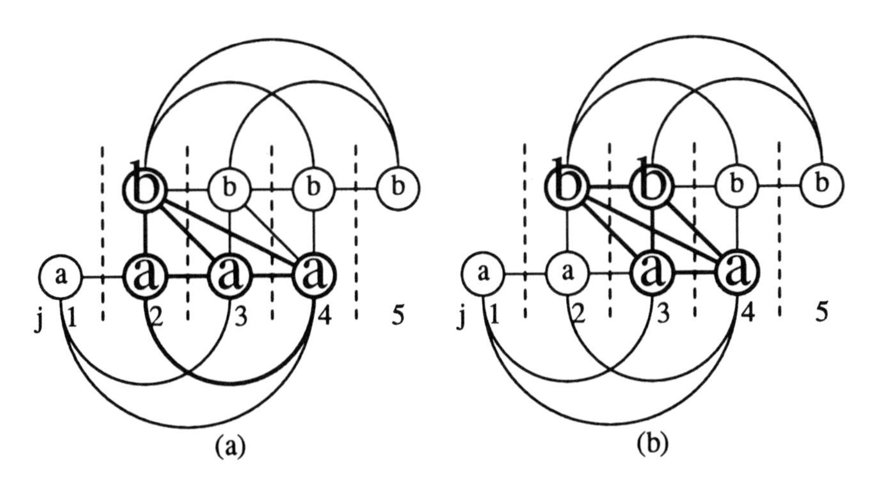

Figure 6.1. Node packing graph for 2 code operations ($a <\!\bullet b$), showing in bold a clique facet for j=2 in (a) and j=3 in (b) of inequality (5).

A different formulation of the one-dimensional precedence constraint formulation was presented in (Baker, 1974, Lee, 1989) as:

$$\sum_{j \varepsilon R(k_2)} (j) x_{j,} k_2 - \sum_{j \varepsilon R(k_1)} (j) x_{j,} k_1 \le -1, \ \forall k_2 <\!\bullet k_1. \tag{6.5*}$$

Even though the set of integer feasible solutions are the same, the formulation (6.5) is tighter than (6.5*) and the proof is given below. To the best of our knowledge inequality (6.5) has not previously been presented in the literature. As we will show in chapter 10 the tightness property is far more important with respect to solving the IP more efficiently than the number of constraints generated (which for 6.5* is less than for 6.5).

To Prove: (6.5) is Tighter Than (6.5*).

Let P^5 represent the scheduling polytope whose constraint set is only generated by (6.5) and similarly for P^{5^*}. This is equivalent to showing that $P^5 \subset P^{5^*}$.

1) $P^5 \neq P^{5^*}$

Proof: Consider the following fractional solution to the scheduling problem $a,b \mid a<\bullet b$, where the upper bound on the number of control steps is 4.

k	j=1	j=2	j=3	j=4
a	.9	0	.1	
b		.5	.5	0

This solution is violated by (6.5)= $(x_{3,a}+x_{3,b}+x_{2,b})=(.1)+(.5)+(.5)=1.1>1$

and feasible for (6.5*)= $\sum_{j=1}^{3} jx_{j,a} - \sum_{j=2}^{4} jx_{j,b} = 1(.9)+3(.1)-2(.5)-3(.5)=-1.3 \leq -1$. Therefore $P^5 \neq P^{5^*}$ QED.

2) $P^5 \Rightarrow P^{5^*}$

Proof:

$$
\begin{array}{lll}
 & x_{3,a}+x_{2,a}+x_{1,a} & =1 \\
 & x_{3,a}+x_{2,a} & \leq 1-x_{2,b} \\
 & x_{3,a} & \leq 1-x_{2,b}-x_{3,b} \\
 & 0 & =1-x_{2,b}-x_{3,b}-x_{4,b} \\
\hline
\text{therefore} & x_{1,a}+2\,x_{2,a}+3\,x_{3,a} & \leq 4-3\,x_{2,b}-2\,x_{3,b}-x_{4,b}
\end{array}
$$

now we have to derive (6.5*): $x_{1,a}+2\,x_{2,a}+3\,x_{3,a}$

$-2\,x_{2,b}-3\,x_{3,b}-4\,x_{4,b}\leq -1$

lhs of (6.5*): $\leq 4-3\,x_{2,b}-2\,x_{3,b}-x_{4,b}-2\,x_{2,b}-3\,x_{3,b}-4\,x_{4,b}$ (by substitution)

$$=4-5\,(x_{2,b}+x_{3,b}+x_{4,b})=-1.$$

Since we have shown that $P^5 \Rightarrow P^{5^*}$ and $P^5 \neq P^{5^*}$ therefore we have proved that $P^5 \subset P^{5^*}$.

End of Proof.

Thus (6.5) is a tighter formulation of the precedence constraint than (6.5*). Thus improvements in IP solution efficiency and better bounds on variables are expected with (6.5) since it is a tighter formulation (Nemhauser, 1988) On further analysis of the graph in figure 6.1, one can see that this graph is strictly characterized by cliques completely generated by (6.1) and (6.5), and G has no chordless odd cycles. Thus G is a perfect graph (by definition) and its integral polytope is completely characterized by inequalities (6.1) and (6.5). For tree structured DAGs, we can prove that the node packing graph generated by (6.1) and (6.5) is a perfect graph by proving that the graph is triangulated (and therefore perfect) using the algorithm in (Golumbic, 1980) We can further show that (6.1) and (6.5) generate all cliques and therefore represent integral polyhedrons for the scheduling problem.

To illustrate how the model changes as we add one more dimension of complexity (functional unit allocation), consider three code operations $\{a,b,c \mid a<\bullet c, b<\bullet c, a,b,c \varepsilon Op(1,1)\}$, with an upper bound of 3 control steps and an upper bound of 2 functional units. The node packing graph for this example is given in figure 6.2. Clearly constraint (6.1) generates cliques and therefore corresponds to integral facets. Constraint (6.2) shown in bold in figure 6.2a) is a clique $\forall$ i, j=1 (by definition), however

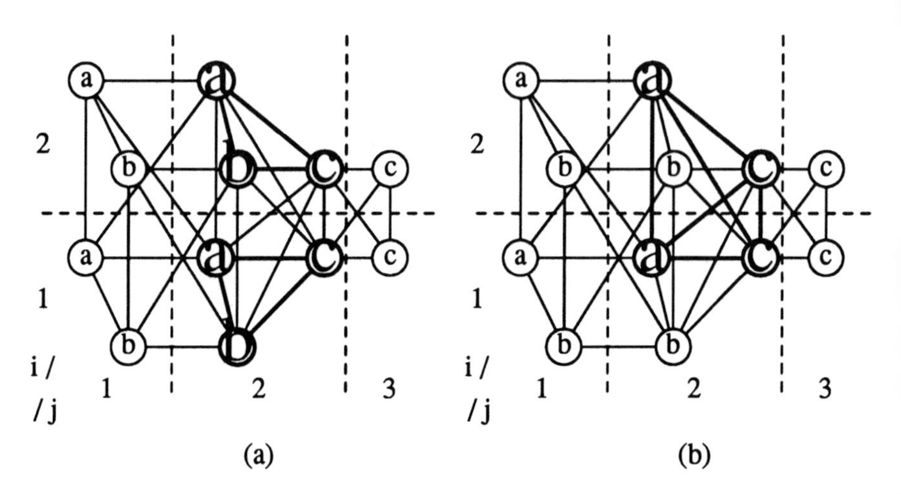

Figure 6.2. Node packing graph for 3 code operations $\{a,b,c \mid a<\bullet c, b<\bullet c\}$ showing in bold inequality (6.2) in a) and inequality (6.6) in b).

it is not a maximal complete subgraph $\forall$ i, j=2. The additional vertex, $x_{2,2,c}$ is required $\forall$ i,j=2 in order to form a clique facet. Thus some generalized inequalities, such as (6.2), may or may not generate facets in a specific algorithmic implementation. Constraint (6.5) for the two-dimensional model can be modified by replacing $x_{j,k}$ with $\sum_i x_{i,j,k}$ as shown in inequality (6.6). Inequality (3) is now redundant and can be removed from the model and replaced with inequality (6.6), which was proven to be a facet of the one-dimensional model. Inequality (6.6) also generates clique facets for the two-dimensional example shown in figure 6.2b). Thus some facets of lower dimensional models, such as (6.5), remain facets when higher dimensions are added (6.6).

$$\sum_{i}\Big(\sum_{\substack{j_1 \le j+C_Z-1 \\ j_1 \varepsilon R(k_1)}} x_{i,j_1,k_1} + \sum_{\substack{j \le j_2 \\ j_2 \varepsilon R(k_2)}} x_{i,j_2,k_2}\Big) \le 1, \tag{6.6}$$

$$\forall k_2 <\bullet\, k_1 \,,\, j \,\varepsilon\, R(k_1) \cap R(k_2) \,,$$

Finally constraint (6.4) can be relaxed to the following constraint (6.7) since we now have some integral facets and can solve using linear programming.

$$0 \le x_{i,j,k} \le 1 \tag{6.7}$$

It is also interesting to note that there are only two additional (lifted odd cycle) facets [3], not given or used in our model (compared with nine clique inequalities) , which are needed to completely characterize the integral polytope for the three code operation example in figure 6.2. Thus the general model for problem 1, simultaneous scheduling and functional unit allocation, consists of variables $x_{i,j,k}$ with constraints (6.1), (6.2), (6.6), and (6.7).

6.2 COST FUNCTIONS

The cost (or objective) function for the IP can be formulated as any linear or piecewise linear function of the variables. An example of the two types of cost functions are the following:

---

3 The additional (lifted odd cycle) facets are $x_{2,1,a}+x_{2,1,b}+\sum_{i,k} x_{i,2,k} \le 2$ and $x_{1,1,a}+x_{1,1,b}+\sum_{i,k} x_{i,2,k} \le 2$.

$$Minimize \sum_{i^0}^{i^1} \sum_{j^0}^{j^1} \sum_{k} (x_{i,j,k} f(i,j,k))$$

$$Maximize \sum_{i^0}^{i^1} \sum_{j^0}^{j^1} \sum_{k} x_{i,j,k}$$

In the first objective function, f(i,j,k), can be a linear or piecewise linear cost function. For example f(i,j,k) = c_fu(i) +c_time(j) can be used to explore the tradeoff of resources. The second objective function is called the feasibility cost function where all resources are fixed and the optimizer determines whether a synthesized hardware solution exists using these bounds.

Many other forms are possible including an area and delay model similar to (Devadas, 1989) which requires additional variables. Two alternative formulations can be used for an area delay cost function. Both will be discussed below.

The first formulation can be used for linear (or very simple piecewise linear) area delay cost functions and introduces the integer variable I_t, where I_t= the number of functional units of type t. The variable I_t can be formulated with the constraint shown below to minimize an area cost function $\sum_t area(t) I_t$, where $area(t)$ is the area of one functional unit of type t. An extension for simple piecewise linear area cost functions is given in chapter 7 section 7.5.

$$\sum_{k \varepsilon t} \sum_{i \varepsilon t} x_{i,j,k} \leq I_t, \ \forall j.$$

The second formulation can be used for piecewise linear cost functions. It uses variables z_{i_t}, where z_{i_t}=1 represents the allocation of one (the ith) functional unit of type t. The activation of continuous variables defined in chapter 4 to threshold the summation of partial indices of the variables is used to formulate the z variable (see inequality given below). The piecewise linear cost function is $\sum_{i_t} z_{i_t}\, a(i_t)$, where $a(i_t)$ is the cost of one (the ith) functional unit of type t. Since the inequality introduced by the binary variable z is not a node packing inequality, the problem is no longer strictly a node packing problem.

$$z_{i_t} \varepsilon \{0,1\}, \;\; z_{i_t} \leq \sum_{j,k} x_{i_t,j,k}, \;\; z_{i_t} \geq (\sum_{j,k} x_{i_t,j,k})/K, \;\; \forall i_t.$$

6.3 FUNCTIONAL UNIT TYPE SELECTION

The structured OASIC model can simultaneously select the type of functional units to optimize the particular cost functions. For example it is possible that for a specific area-delay cost function a pipelined multiplier provides a more optimal implementation as opposed to a multicycled multiplier. We can use a different general constraint on each mapping of code operation to type of functional unit. For example to select a 3 cycle (t=1) or 2 cycle (t=2) multiplying functional unit we generate two constraints using inequality (6.6). First we replace $\sum_{i}$ with $\sum_{i_t}$ for $k_2 <\bullet k_1$, $k_2 \varepsilon t$, of inequality (6.6) and use C_2=3 for t=1 and C_2=2 for t=2. Inequality (6.2) would also be used more than once for each code operation in the same manner.

Constraints can also be formulated for chaining operations. For example the designer could optimally determine whether it is advantageous to select (simultaneously with allocation and scheduling) a functional unit that chains operations. A new type of functional unit (t=3) is created (to represent the chained operations). For t=3 and $k_1 <\bullet k_2$ where k_1 and k_2 ε t=3, a fixed timing constraint between k_1 and k_2 is used in place of inequality (6.6). The fixed timing constraint (to be discussed in detail in chapter 9) essentially sets $x_{i_3,j,k_1}=x_{i_3,j,k_2}, \forall i_3,j$. For example let us assume the functional unit solves ((c*d) + (a+b)) in one cstep by chaining the two additions and one multiplication together. Then the fixed timing constraints for $k_+ <\bullet k_+, k_* <\bullet k_+$ is formulated. Additionally we need the following constraints $\sum_{\substack{k \\ k\varepsilon+}} x_{i_3,j,k} \leq 2, \forall i_3,j$ and $\sum_{\substack{k \\ k\varepsilon*}} x_{i_3,j,k} \leq 1, \forall i_3,j$ to ensure only two adders and one multiplier are allocated. This formulation also allows any addition or chaining of two additions or any multiplication or ((c*d) + e) or ((c*d) + (a+b)) to share the functional unit.

The OASIC model for simultaneous scheduling and allocation and selection of functional units was presented in this chapter. The node packing characteristics, introduced in chapter 4, were used to extract facets of the polytope. The next model to be presented in chapter 7 directly supports area-delay cost functions, which are a good vehicle for design exploration and an important part of the design methodology as discussed in chapter 3 and 5. In the OASIC model of the next chapter the very tight (node packing) precedence constraints, presented in this chapter, are kept and unstructured constraints are used for allocating the hardware resources.

7.

OASIC: AREA-DELAY CONSTRAINED ARCHITECTURAL SYNTHESIS

The major difference between the model described in this chapter and the previous structured model in chapter 6 is that we cannot simultaneously optimize the selection of types of functional units without the introduction of nonlinear inequalities†. In other words the mapping of operations to types of functional units must be a one to one (or many to one) mapping.

The area-delay OASIC model directly supports area-delay cost functions without disjunctive constraints, the latter required by the OASIC model in chapter 6. The area-delay model in general trades off structure

† For example if $\lambda=1$ select a two cycle multiplier, *, else ($\lambda=0$) select a pipelined multiplier, *pl, inequality (7.3) becomes, $\sum_{k_*} x_{j,k_*} + \lambda \sum_{k_*} x_{j-1,k_*} \leq I_*$

for size (in variables and equations). Although the constraints are unstructured we have identified application specific cases where we can lift some constraints and employed the use of knapsack inequalities to further tighten the model.

Preprocessing now involves selection of functional unit types. Types of functional units can be selected by the designer by hand or by using the previous model (chapter 6) to simultaneously schedule and select and allocate functional units. The asap and alap preprocessing described in 3.3.1 is also required for the area-delay model.

Other preprocessing unique to this model involves edge reduction of the DAG for register allocation constraint and the identification of special operations to be defined in section 7.6. The later is used for tightening constraints. The edge reduction algorithm will be described in this chapter. Finally after lower bounds are computed knapsack inequalities, defined in chapter 4, can be automatically generated to improve these bounds. An automatic algorithm for extracting knapsack inequalities is explained in (Crowder, 1983) . We will briefly discuss the variation for our particular application, architectural synthesis.

7.1 THE PRECEDENCE CONSTRAINED SCHEDULING MODEL

The operation assignment constraint, (7.1), ensures that each operation will be assigned to one cstep. The precedence constraint, (7.2), prevents an operation k_2 from being scheduled after operation k_1 whenever there is a partial order between these operations such that $k_2 <\bullet k_1$. This constraint is the same as the precedence constraint (6.5) presented in chapter 6.2.

$$\sum_{j \varepsilon R(k)} x_{j,k} = 1, \forall k. \tag{7.1}$$

$$\sum_{j_1 \le j, j_1 \varepsilon R(k_1)} x_{j_1,k_1} + \sum_{j-(C_2-1) \le j_2, j_2 \varepsilon R(k_2)} x_{j_2,k_2} \le 1, \tag{7.2}$$

$$\forall k_2 <\bullet k_1,\ j \varepsilon R(k_1) \cap (R(k_2)+C_2-1).$$

7.2 FUNCTIONAL UNIT ALLOCATION

The functional unit constraint, (7.3), ensures that no more than I_t functional units of type t (Lee, 1989, Baker, 1974) will be required in the solution.

$$\sum_{\substack{k \varepsilon t \\ j \varepsilon R(k)}} \sum_{j1=j}^{j1=j+(L-1)} x_{j1,k} \le I_t,\ \forall t, j. \tag{7.3}$$

7.3 REGISTER ALLOCATION

The register allocation constraint ensures that there are no more than R variables whose lifetimes overlap at any cstep. A variable lifetime can be represented by a (lifetime-defining) edge $(k <\bullet k_l)$ between the defining operation, k, and the operation which last used the variable,k_l. However in many algorithms each variable may be input to more than one code operation $(k <\bullet k_e | e > 1)$, thus it is difficult to determine which k_e should be the lifetime-defining edge. Two properties, transitivity and alap analysis, can be used to decrease the number of edges we must consider for representing a variable lifetime. For example in figure 7.1a), (e=1,2) transitivity requires $(k_1 <\bullet k_2) \Rightarrow (k_2 = k_l)$ and in figure 7.1b) alap analysis requires $(asap(k_2) \ge alap(k_1)) \Rightarrow (k_2 = k_l)$. This preprocessing can be done very fast and is outlined below:

Edge Reduction Algorithm

Given the node-node adjacency matrix $A_{i,j}$, where $a_{i,j}=1 \Leftrightarrow i <\bullet j$, where i,j are code operations. We will use this matrix to represent the DAG (or input algorithm). First a path matrix, T, is calculated. Secondly the asap/alap tables for each code operation and the path matrix are used to delete edges that cannot represent lifetimes.

1) Compute matrix $T_{i,j}$, where $t_{i,j}=1 \Leftrightarrow \exists$ a path from code operation i to code operation j in the DAG. A depth first search of the DAG will find T. The general structure and pseudocode notation of this algorithm was taken from (Golumbic, 1980) .

```
Procedure DFSEARCH(v,L):
    begin
    mark v "visited" & set v ε L
    for each w ε Adj(v) do
        if w is marked "unvisited" then
            begin
            set t_{v,w}=1 ∀ v ε L.
            DFSEARCH(w,L);
            end
        else if w is "visited" &&
            t_{v,w} == 0 then
            begin
            set t_{v,w}=1 ∀ v ε L
            t_{v,k}=t_{w,k} ∀ k
            end
    end
```

2) Compute matrix $L_{i,j}$, where $l_{i,j}=1 \Leftrightarrow$ edge i,j of the DAG cannot be eliminated by transitivity or alap analysis. Initially set L = A, and then eliminate entries $\forall$ i,

$\forall\ j_1, j_2 \mid j_1 \neq j_2,\ a_{i,j_1} = a_{i,j_2} = 1.$

$if((t_{j_1,j_2}=1) \vee (asap(j_2) \geq alap(j_1))) \Rightarrow l_{i,j_1} = 0.$

The L matrix is used to generated the register constraints since each entry represents a possible lifetime defining edge.

We will now describe how the register allocation constraint (7.4) can ensure no more than R registers are allocated even with multiple edges representing a variable's lifetime. The following terminology is used: (a) An arc, $k_n <\bullet k_e$ (whose head is k_n and tail is k_e), is said to *cross* cstep j if and only if $R(k_n) \cap \{0,1,...,j\} \neq 0$ and $R(k_e) \cap \{j+1, j+2, ..., Te\} \neq 0$; (b) e(n) = the number of arcs $(k_n <\bullet k_e, e \geq 1)$, with head k_n that cross at j (e(n) ≤ e). For the general case where $e(n) \geq 1 \forall n$, constraint (7.4) is generated, $\prod_n e(n)$ times per cstep, for all maximal sets of arcs that cross j such that no two arcs in a set have the same head. For example if only one head (k_i) has e multiple arcs $(k_i <\bullet k_j, j=1,...,e.)$ that cross at j $(e(i)=e)$, and the rest of the arcs have unique heads $(e(n)=1 \forall n \neq i)$, then (7.4) is generated e times (once for each k_j). In practise the number of constraints will not be a significant problem, because 1) the computation time for IP problems is not highly sensitive to the number of constraints (Nemhauser, 1988) and 2) most algorithms will have small values of $\prod_n e(n)$ which intersect at the same cstep. The register allocation constraint (7.4), calculates two times the number of cut edges at each cstep, by dividing time and operations into four quadrants as shown in figure 7.2. For all $k_e \in K_e^n = \{k_e \mid k_n <\bullet k_e, e \geq 1\}$ or $\{k_l\}$ means that the constraint (7.4) is generated for all maximal sets of arcs that cross j such that no two arcs in a set have the same head.

$$\sum_{\substack{k_n \\ k_n <\bullet k_e}} \Big(\sum_{\substack{j1 \le j-(C_n-1) \\ j1 \varepsilon R(k_n)}} x_{j1,k_n} + \sum_{\substack{j2>j \\ j2 \varepsilon R(k_e)}} x_{j2,k_e} - \sum_{\substack{j3\le j \\ j3 \varepsilon R(k_e)}} x_{j3,k_e} \qquad (7.4)$$

$$- \sum_{\substack{j4 > j-(C_n-1) \\ j4 \varepsilon R(k_n)}} x_{j4,k_n} \Big) \le 2\,R, \forall j, k_e \,\varepsilon\, K_e^n$$

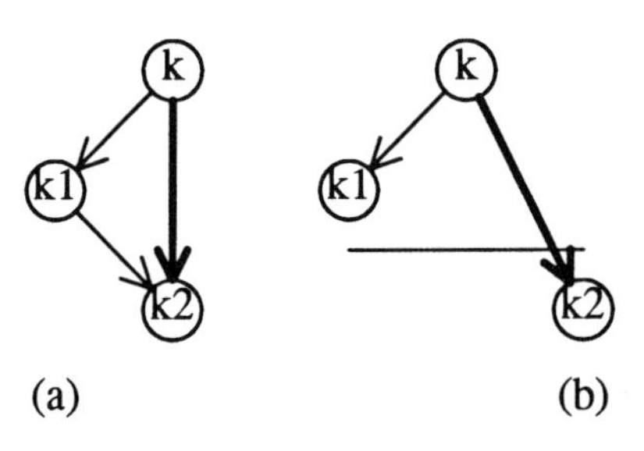

Figure 7.1. Lifetime defining edges for the variable output from k are shown in bold using (a) transitivity analysis and (b) alap analysis ($alap(k_1) \le asap(k_2)$).

7.4 BUS ALLOCATION

Since we are interested in obtaining an exact measure of the number of busses (defined in chapter 2) of an architecture, we define the number of parallel data transfers (*pdt*) as the maximum number of data transfers that occur at one time (counting transfers with distinct sources as the number of destinations as discussed in chapters 6 and 3) unlike previous (Devadas, 1989, Paulin, 1989) definitions. We constrain each hardware unit (register or functional unit) to have only one bus per input, unlike other heuristics (Huang, 1990) which cannot estimate additional multiplexors required later in the synthesis process.

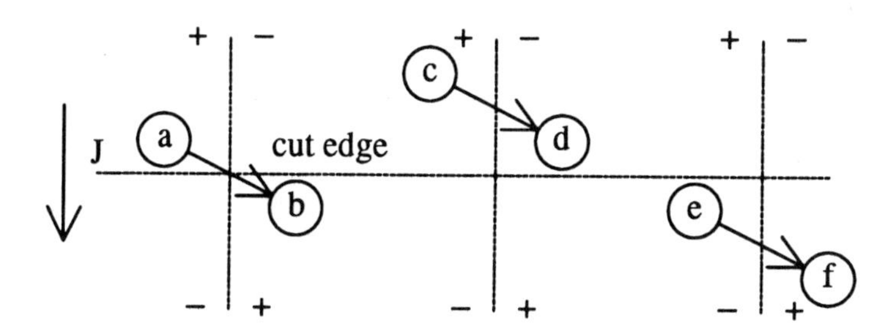

Figure 7.2. Register allocation constraint illustrated with one cut edge, $a<\bullet b$, at J (the other two edges cancel out because each edge has nodes in a '+' and '-' quadrant).

As discussed in chapter 3 bus allocation even with fixed schedules is difficult due to the possibility of global data broadcasts. We will outline our approach to handling data broadcasts, that ensure an exact number of busses is calculated. In our model the user must identify operations which will be transmitted data by global data broadcasts. Data broadcasts can therefore be modeled using fixed timing constraints (Gebotys, 1991c) on all pairs of destination operations and selecting one of the destination operations to contribute to the bus count. For the remainder of this chapter we will assume that these constraints have been incorporated and will not address data broadcasts further. Next the formulation is presented for simultaneous scheduling and bus allocation.

The bus allocation constraint, (7.5), ensures that at each cstep no more than B busses are required to transfer data between functional units and registers. An additional constraint $(In(t))I_t \leq B, \forall t$, or $(In(t)+Out(t))I_t \leq B, \forall t \varepsilon op(1,1)$ is also used in the OASIC model to decrease the size of the search space. We will now show that this defined number of parallel data transfers is exactly equal to the number of busses in an optimal architectural solution for 1) module allocation, 2) allocation of at most 2 types of functional units and 3) other cases. This is analogous to channel routing, where the left edge algorithm is guaranteed to find a route (architecture) that uses no more than B tracks (busses), calculated from the channel density (*pdt*), even if we require at most two sets (of data transfers to each of two inputs of a functional unit) of nets to be placed on distinct tracks (one distinct bus per input to each input of a functional unit). Previous research has not addressed this problem. As discussed in chapter 3, other synthesizers allocate busses after functional unit allocation.

$$\sum_{\substack{k \\ j \varepsilon R(k)}} (In(k)) x_{j,k} + \sum_{\substack{k_1 \\ j1 \varepsilon R(k_1) \\ j1 = j - (C_1 - 1)}} (Out(k_1)) x_{j1,k_1} \leq B, \forall j. \quad (7.5)$$

For t=3 the problem in the worst case requires $\lfloor pdt/3 \rfloor$ additional busses and is NP complete. For example assume we have three (single cycled) functional units of different types (t=3) and all pairs of functional units are scheduled in parallel (but all 3 functional units are never scheduled at once). For example in figure 7.3a) the scheduled DAG is on the left and one possible architecture, with 3 types of functional units (+,*,-), is shown on the right. The *pdt* would be calculated as (2*2+2=) 6, but (3*2+2=) 8 busses are required since each functional unit must have only one bus per input. Analogous to segmented channel routing, each of the three sets of data transfers into the three types of functional units must be placed on a separate segment (or input bus to that type of functional unit). However the *pdt* is exact when not all pairs are scheduled at the same time (more than one type of functional unit can share a input bus with another different type of functional unit) or when all 3 are scheduled at least once in parallel or many other cases such as figure 7.3b). Nevertheless since many DSP algorithms have only two types of functional units and in many practical applications the functional units have high utilization, the *pdt* will often be exactly equal to the number of busses for $t \geq 3$.

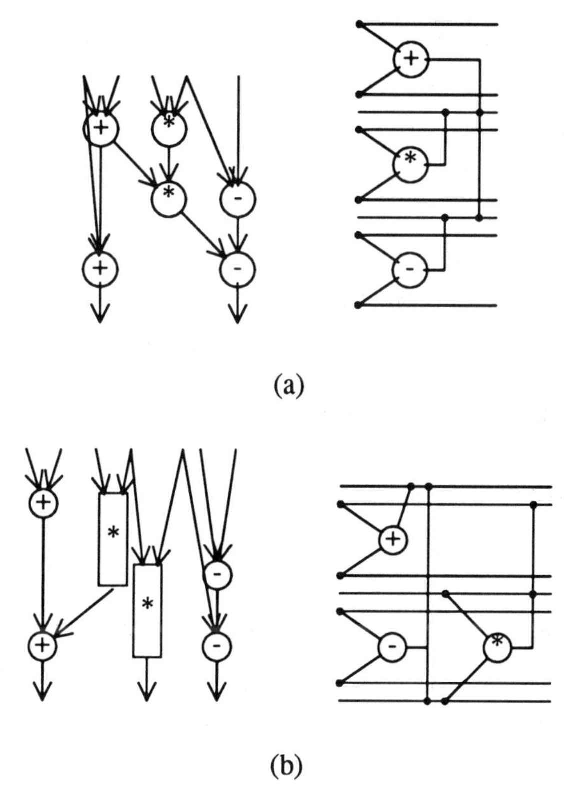

Figure 7.3. *Pdt* for $t=3$ in (a) is not equal to B ($pdt=6$, B=8) and in (b) *pdt* is equal to B (pdt=B=7).

We will now present proof that an architectural solution with B busses, R registers, and I_t functional units of type t ($t \leq 2$) is always guaranteed to exist. We will prove this in two steps for the case where code operations are single cycle and therefore $B^{in} \cap B^{out} = \varnothing$, (where $B^{in}(B^{out})$ is the set of busses input to (output from) functional units,

$|B^{in}|+|B^{out}|=B$).

First we must show that an architecture exists with R registers, such that each register can have only one input bus chosen from B^{out} busses. This problem is independent of I_t functional units since each functional unit can output data to any bus (B^{out}) in theory. Each set of variables (a set has variables with nonoverlapping lifetimes ie. a register) can be merged (onto the same input bus) if no two variables have the same definition times. We can always obtain $|B^{out}|$ merged sets, by swapping groups of variables between sets, and therefore both a minimum number of registers and busses is possible. In other words, analogous to channel routing this partitions R tracks into B^{out} busses or $B^{out} = \underset{(\forall j)}{Max}$ { # of variables defined at cstep j} and R = $\sum_{l} \underset{(\forall j)}{Max}$ {# of overlapped lifetimes of variables assigned to bus l at cstep j}.

Secondly we can show that an architecture with $|B^{in}|$ busses (where each input of a functional unit can only be assigned to one input bus) and I_t functional units can be guaranteed. This problem is independent of R since each register can output variables to any bus (B^{in}) in theory. The first observation is that input busses can only be shared between functional units of different types (in order to guarantee a minimum of I_t functional units). For example if two ALUs can share input busses then they are never scheduled at the same time and therefore only one ALU is needed. Since we do not consider global data broadcasts here, functional units can share both input busses or none. We can now form a complete bipartite graph, $K_{X,Y}$, where the X vertices represent functional units of type t=1, and the Y vertices represent functional units of type t=2. For each cstep of the schedule, we assign code operations to functional unit vertices of G and delete all edges between the assigned vertices. At each successive cstep we assign code operations to previously assigned functional unit vertices and when this is no longer possible we assign them to new functional unit vertices and delete new edges. After all csteps have been exhausted, the cardinality of the maximum matching of the final graph, |M|, is the number of input busses shared by functional units (therefore $|B^{in}|=2(\sum_t I_t - |M|)$).

An extension to this proof (t=2) for multicycle operations can also be done. In this case $B^{in} \cap B^{out} = B^{io} \neq 0$ (therefore whenever B^{io} is being used as an input bus, other functional units cannot output variables to $B^{io} \subset B^{out}$) , however since $B^{out} < B^{in}$ we can use the two separate proofs above to show that a minimum R and I_t can exist and pdt guarantees that B busses are needed. The OASIC model is exact in allocation of all resources ($t \leq 2$) except bus drivers (which cannot be allocated before the binding phase, see Chapter 10). For the first time this provides us with an exact defined relationship between parallel data transfers and the number of busses required in the architecture.

7.5 COST FUNCTIONS

Piecewise linear cost functions, defined in (Devadas, 1989) and for example shown in (7.6) below, are also supported, where $T_e = \sum_j (j-1) x_{j,k_{out}}$. The special operation k_{out} is used when there is more than one operation which outputs a value at the end of the algorithm (or loop). Each of these last operations is defined to precede k_{out}. Then k_{out} is used to define the last cstep of the algorithm or the end the loop. Similar to k_{in}, k_{out} is only used for the partial order and register allocation constraint and it does not participate in the functional unit allocation constraints. Other general piecewise linear functions, where for example the cost per register is different if there exists more than 5 registers are defined in (Nemhauser, 1988) and can be also modeled but require additional binary variables.

$$\sum_{t} c\_fu(t)\, I_t + c\_bus\; B + c\_reg\; R + c\_time\; T_e \tag{7.6}$$

To illustrate a piecewise linear cost function assume we have a cost of 10 per register for the first 5 registers and a cost of 15 per register for the additional registers (Devadas, 1989) after the fifth. The register constraint (7.4) becomes $\alpha x \leq 2(R_1+R_2)$, $R_1^{UB}=5$ and $R_2^{LB}=0$, and part of the cost function becomes $(10\ 15)(R_1\ R_2)^T = c\_reg\ R$.

7.6 APPLICATION SPECIFIC TIGHTENING OF CONSTRAINTS

We additionally use k_{sep} (and k_{out}) of the DAG to tighten functional unit and bus allocation constraints. These operations are present whenever one code operation, k_{sep}, precedes and/or is preceded by (<•>) all other operations. This operation is present at the beginning/end of loops, branches, and algorithms, and sometimes within basic blocks of code (such as the elliptic wave filter see chapter 10). Tightened inequalities for functional unit allocation (7.7) and the bus allocation (7.8) are shown below. In (7.7) if k_{sep} is not of type t then its coefficient is ($I^{LB}{}_t$) and in (7.8) if k_{sep} is not a single cycle operation then its coefficient is $(B^{LB}-In(k_{sep}))$.

$$\sum_{\substack{k \varepsilon t \\ j \varepsilon R(k)}} \sum_{j1=j}^{j1=j+(L-1)} x_{j1,k} + (I_t^{LB}-1)\, x_{j,k_{sep}} \leq I_t,\ \forall t, j \varepsilon R(k_{sep}), k_{sep} \varepsilon t. \tag{7.7}$$

$$\sum_{\substack{k \\ j \varepsilon R(k)}} (In(k))x_{j,k} + (B^{LB} - In(k_{sep}) - Out(k_{sep}))\, x_{j,k_{sep}} \tag{7.8}$$

$$+ \sum_{\substack{k_1 \\ j1 \varepsilon R(k_1) \\ j1 = j-(C_1-1)}} (Out(k_1))x_{j1,k_1} \leq B,\ \forall j \varepsilon R(k_{sep}).$$

Knapsack inequalities of the bus constraint can be used to tighten the OASIC model whenever the coefficients of the x variables in (7.8) or (7.5) are different. This occurs whenever code operations of the DAG have different numbers of input variables or multicycle functional units are used. In the OASIC methodology the (integer rounded) lower bound calculation for the number of busses is fixed and knapsack inequalities are extracted. The relaxed linear program is then resolved to determine if it is still feasible. If it is infeasible then the lower bound on the number of busses is incremented. Chapter 10 will illustrate this procedure and show the cpu speed improvement attained. The number of inputs to operations may also be application dependent, for example if one input of all multiplication operations is obtained from memory then only one input bus needs to be allocated since we can directly connect the other inputs to specific memory.

The register allocation constraint could be converted into a knapsack inequality (by changing the -1 coefficient into +1 coefficient, ie. $x_1 = 1 - x$), however poor results are expected since the variable coefficients are all one. Currently the register allocation constraints are not tightened.

In this chapter we presented the OASIC model for area-delay cost functions. We also described how special operations and the type of functional units can be used to tighten and extract facets of the underlying polytopes. The next chapter will examine how instances of these constraints are used to support algorithmic constructions such as conditional code, loops, and functional pipelining in OASIC.

8.

SUPPORT FOR ALGORITHMIC CONSTRUCTS

We will discuss in this chapter model extensions for conditional code, loops and functional pipelining. In addition their effect on timing constraints will also be covered in relationship to the architectural synthesis model and controller implications. These constructs are supported by both the structured and area-delay optimization models presented in chapter 6 and 7. We will present their formulation using the area-delay model for simplicity. However one can transform these into constructs for the structured model by substituting $\sum_i x_{i,j,k}$ for $x_{j,k}$.

8.1 CONDITIONAL CODE

All the previous inequalities in chapter 6 and 7 apply to basic blocks where operations are not mutually exclusive. We will now address how the previous constraints presented in chapter 6 and 7 can be used to support conditional code. Conditional code is supported by applying constraints (6.3),(6.4), and (6.5) to code operations in separate mutually exclusive code segments. If we let B_i represent basic blocks of code (or straight line code) in the algorithm then we can define a branch as a precedence constraint between blocks. For example blocks of code $B_1 <\bullet B_2$, $B_1 <\bullet B_3$, $B_2 <\bullet B_4$ and $B_3 <\bullet B_4$ can be used to represent the blocks of code in figure 8.1. The code operations in B_2 and B_3 are mutually exclusive. The inequalities for functional unit allocation are given in (8.1). The rest of the inequalities are also applied to these basic blocks. If we assume conditional branches have equal probability then we may have to add further data precedence constraints in order to a) prevent code motion (Ellis, 1986) or b) prevent conditional code operations being scheduled (illegally) before the branch or after the join of the branch.

$$\sum_{\substack{k \varepsilon B_z \\ j \varepsilon R(k)}} x_{i,j,k} \leq 1. \forall i, j, z = 1,2,3,4. \tag{8.1}$$

OASIC can be used to minimize the weighted sum of execution times for all conditional paths. Therefore one can schedule and allocate resources simultaneously with minimizing the execution time on different paths. This is similar to the k_{out} placed at the end of loops and algorithms to measure the last cstep and ensure registers are allocated to hold these output values. For our case at the end of each branch if more than one operation outputs a value we use partial order constraint to ensure it is succeeded by k_{eob} or the end of branch operation. We can then use the scheduled time of k_{eob} in our objective function along with all other end of branch operations to minimize the individual conditional paths.

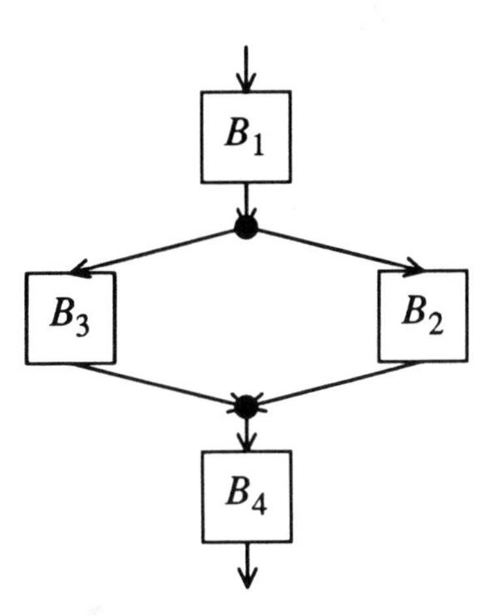

Figure 8.1. Conditional code illustrated with four basic blocks of code operations, where B_2 and B_3 are mutually exclusive.

It is also interesting to note that conditional constructs do not present a problem for register allocation unlike the case for allocating registers for fixed schedules with conditionals. Thus for the first time we can minimize the number of registers in the presence of conditional code. Furthermore edge reduction is not required for cases where an edge originates from one operation (before the branch) and terminates at operations in mutually exclusive basic blocks. Edge reduction is not necessary since the constraint generation is done on a per branch basis so in each inequality only one edge is seen.

Only during functional pipelining (described in chapter 8.3), where the maximum execution time of each conditional path is not fixed, is it necessary to transform the DAG (with conditionals) into an outtree (see chapter 3.4) structure where there are only branches and no joins. However when the execution times of the conditional paths are known the

DAG transformation is not necessary, and time translation can be used to allocate resources from different pipestages.

Timing constraints between operations before conditional branches and operations after conditional branches will not pose problems for scheduling. If conditional paths differ in length then one may have to append empty states to the controller in order to ensure that specific minimum timing constraints are met.

8.2 LOOPS

Loops are easily supported in OASIC using special k_{in} and k_{out} operations which participate in all constraints except (7.3). These operations ensure that output (loop) variables are valid until the end of the (loop) algorithm, and input (loop) variables are valid until their last use inside the algorithm. For example variables input to the loop are represented by $k_{in} <\!\bullet k_1, \forall k_1$ and variables output from the loop are represented by $k_2 <\!\bullet k_{out}, \forall k_2$.

Timing constraints can also be incorporated within or across loops. For a minimum timing constraint, we assume the loop executes a minimum number of times. For a maximum timing constraint, we assume the loop executes a determinate number of times as in (Hayati, 1989) .

8.3 FUNCTIONAL PIPELINING

Functional pipelining for a fixed latency, l, can be incorporated into our model without additional variables. We use the term functional pipelining to refer to executing a number of instances of the input algorithm in parallel but each successive instance is delayed in time. We call the delay in time, for each pair of successive instances, the latency. At most p instances (or pipestages) of the input algorithm are executing in parallel at one time (at one clock period, or at one cstep). Furthermore each instance of the algorithm has the identical schedule of code operations except it is delayed by the latency.

We assume Te = J^{UB} and define $\lceil T_e/l \rceil$=p pipestages, and replace $\sum_{k \mid j \varepsilon R(k)} x_{j,k}$ of (7.3),(7.4),(7.5) with $\sum_{n=1}^{n=p} \sum_{\substack{k \\ (j+nl) \varepsilon R(k)}} x_{j+nl,k}$ where addition $(j+nl)$ is modulo l. Only constraints for l csteps need to be generated as shown in figure 8.2(a), where A through E represents sets of code operations scheduled over each successive group of l csteps.

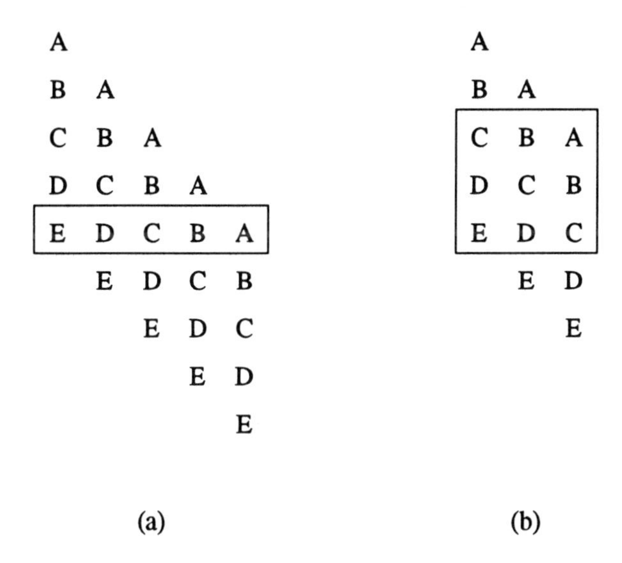

Figure 8.2 Pipelining constructs showing in boxes the number of csteps required to generate OASIC allocation constraints (l in (a) and $3l$ in (b)).

A variation of the functional pipelining can also be used if only z pipestages with latency l are used. In this case $z<p$, only zl csteps are used to generate constraints representing the period from zl to J^{UB} csteps as shown in figure 8.2(b), where J^{UB} is the upper bound on the number of csteps for one pipestage. In this example three pipestages are used each with a latency equivalent to the number of csteps in A.

We can also support functional pipelining in the presence of conditional code. This construction generates the functional pipelining constraints for each combination of conditional branches which are possible over all pipestages at each cstep.

General algorithmic constructs were presented in this chapter for the OASIC model. In the next chapter we examine the special constraints for interfacing to analog, asynchronous or other processes. The next chapter concludes the discussion on constraint formulation for OASIC.

9.

INTERFACE CONSTRAINTS

Timing constraints can be used to represent interfaces with external processes or local constraints between pairs or groups of code operations. Five categories of timing constraints are analyzed below. We will assume timing constraints are given in terms of the clock period. Otherwise we can convert using the following: we use $\lceil t/t_c \rceil = T$ for minimum timing constraints and $\lfloor t/ t_c \rfloor = T$ for maximum timing constraints, where t_c is the period of the clock (equal to one cstep) and t is the real time constraint given by the specification. We will investigate the following types of timing constraints for architectural synthesis:

- Fixed Timing
- Minimum Timing

- Maximum Timing
- Unknown Timing
- Special Timing Constraints

Fixed timing constraints can be used to represent an analog interface. An example is the sampling of an analog signal (using an analog to digital converter), where successive digital data is input to the architecture to be synthesized after every T csteps. The minimum and maximum timing constraints could also specify an interface with some external process where after at least (or at most) T units of time (since a code operation output data to the external process), data will be available, from the external process, in an input register. Asynchronous interfaces occur when one deals with unknown delays. An unknown, bounded timing constraint is an example of an asynchronous type of interface. Bounded means that an interval of time (lower bound and upper bound) when data may be received for input to a code operation is known. For example a designer could know that anywhere from cstep 5 to cstep 10 the data will be input. A bounded data dependent loop also represents a bounded unknown delay. Unbounded unknown delays, such as data dependent loops or synchronizers, will also be covered. In all examples below, T is the time constraint value. We use the notation time(k, k_2) $<\geq$ T to represent the (minimum >, maximum <, or fixed =) time constraint between the two operations. Combinations of these constraints are also possible as discussed in section 9.5.

9.1 GENERAL INTERFACE: MINIMUM AND MAXIMUM TIMING CONSTRAINTS

Minimum and maximum timing constraints can be easily incorporated and have the same form as the precedence constraint and therefore are very tight. The minimum and maximum timing constraints between the scheduled csteps of two code operations can be represented by the inequalities (9.1) and (9.2). This is equivalent to setting $C_2=T$ in inequality (6.6) of chapter 6.

$$\sum_{j>j_1} x_{k_1,j} + \sum_{j\leq j_1+T} x_{k_2,j} \leq 1. \quad \forall j_1, time(k_1,k_2) \geq T. \tag{9.1}$$

$$\sum_{j<j_1} x_{k_1,j} + \sum_{j\geq j_1+T} x_{k_2,j} \leq 1. \quad \forall j_1, time(k_1,k_2) \leq T. \tag{9.2}$$

9.2 ANALOG INTERFACE: FIXED TIMING CONSTRAINT

Interface with analog processes can be modeled using fixed timing constraints, between operations which successively output data to DAC or receive input data from ADC. A fixed timing constraint of T ($T\geq 0$) between two operations can also be defined as the scheduled time for operation 2 is T cycles after operation 1. This can be represented by the following equality in our assignment model, $x_{k_1,j}=x_{k_2,(j+T)}$, $\forall$ j. However the tighter formulations of these constraint are the following node packing facets of the scheduling problem shown in (9.3).

$$x_{k_1,j_1} + \sum_{j\neq j_1+T} x_{k_2,j} \leq 1. \quad x_{k_2,j_1} + \sum_{j\neq j_1+T} x_{k_1,j} \leq 1. \tag{9.3}$$

$$\forall j_1, time(k_1,k_2) = T$$

The above facets are used along with minimum and maximum timing constraints, time$(k_1,k_2) \leq$ T and time$(k_1,k_2) \leq$ T that were presented in the previous section 9.1. Figure 9.1. gives an example of these facets for T=0. A similar comparison can be made with Baker's (Baker, 1974) and Lee etal.'s (Lee, 1989) formulation as done in chapter 6.1, to show that our formulations are tighter than theirs.

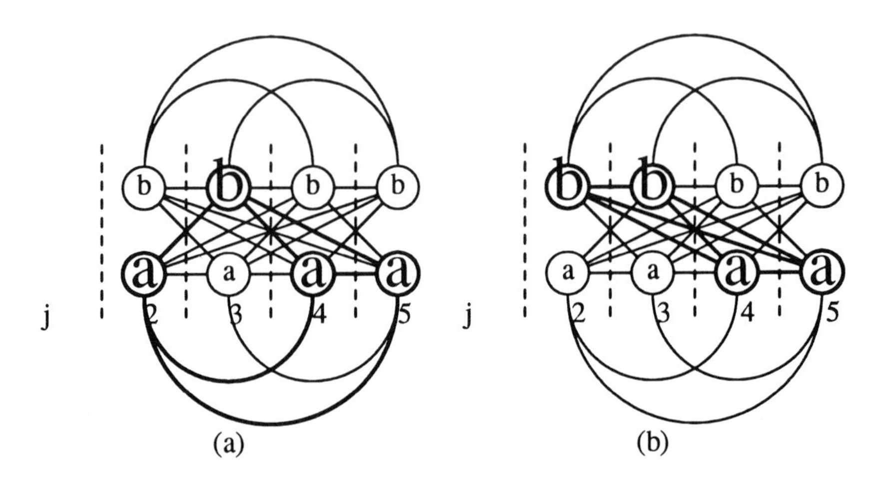

Figure 9.1. Node packing graph for fixed timing constraint between 2 code operations *time*(a,b)=T=0. (a) shows the fixed timing constraint and (b) shows the minimum/maximum constraints, both are facets of the underlying polytope.

9.3 ASYNCHRONOUS INTERFACE

Interface with asynchronous processes can be modeled as bounded (or unbounded) unknown timing constraints. We will address different approaches to the bounded asynchronous interface in this section. Let $t_{\max}$ represent the latest possible time for the data to arrive (upper bound) and $t_{\min}$ the earliest possible time for the data to arrive (lower bound). Then $p = (t_{\max} - t_{\min})$ is the number of control steps when the data may arrive. We assume each control step has equal probability of arriving and we wish to minimize the overall algorithm execution time, the control store and total resources. We will use the terminology defined in chapter 2 to describe the partition of operations resulting from the asynchronous interface, ie. interface dependent operations and interface independent operations. For example consider a code operation, k_{out}, which outputs data to an asynchronous operation, K_a. Operation K_a performs data dependent processing and returns its output data after some indeterminate (and bounded) number of csteps. Assume that the bound on the indeterminate number of csteps is p. We can expect to receive data from K_a at cstep = $j_r \mid asap(k_{out}) + C_{out} + d \le j_r \le alap(k_{out}) + C_{out} - 1 + d + p$, where d + 1 (or (d+p)) is the minimum (or maximum) processing time of K_a. The following five solutions were considered below.

a) Trivial Controller Wait State

b) Partition Resources: Tradeoff Hardware for Control

c) Maximum Resource Sharing: Tradeoff Control for Hardware

d) Mutually Exclusive Pipelining

e) As Late as Possible Approach: Tradeoff Execution time for Hardware & Control

Each case will be discussed below with respect to synthesis complexity, hardware versus controller area (number of words to store) and execution time tradeoffs.

a) Trivial Controller Wait State

In the trivial-controller-wait-state the controller waits from cstep=$asap(k_{out}) + C_{out} + d$ until data from K_a arrives. For architectural synthesis we can synthesize all code operations assuming that the data from the asynchronous operation is ready at $asap(k_{out}) + C_{out} + d$. In OASIC a minimum timing constraint ($C_{out} + d$) is placed between k_{out} and the asynchronous operation K_a (which is modeled in OASIC as a single cycle operation).

b) Partition Resources: Tradeoff Hardware for Control

In this scenario we can assume that while part of the interface dependent algorithm is waiting for the input data, the remaining interface independent algorithm is executing on the architecture until it becomes interface dependent, as illustrated in chapter 2, figure 2.4. We assume that the interface dependent code operations are executed on hardware that cannot be shared by interface independent code operations. Figure 9.2a) shows the original data flow graph, where K_a is the asynchronous operation. The partitioned schedule for two separate controllers is in figure 9.2b). This can be estimated as a controller size of 6, since 6 words of control (or control states) are required. In this case data path synthesis can be partitioned into two routines, in (b), to be synthesized on separate hardware.

c) Maximum Resource Sharing: Tradeoff Control for Hardware

Solution (b) may be very inefficient if a large number of functional units are required for the interface dependent algorithm. Therefore in this section we investigate sharing the hardware between both algorithms. This problem is now defined as scheduling and allocating hardware for conditional code (representing interface dependent code) originating from each successive cstep within the bounded interval p. We can use p mutually exclusive branches (b=1,...,p) of interface dependent code operations where each branch starts at $d+b$ csteps after k_{out} outputs data $k_{out}<\bullet K_a$ (using fixed timing constraints for b=1). This is shown in figure 9.3 (a) and (b). It can be seen that the controller in (b) may become very large because we allow independent scheduling of each branch to optimize the architecture, however only one schedule for the interface independent code is used. Separate variables are used to represent the scheduling of code operations in each mutually exclusive branch.

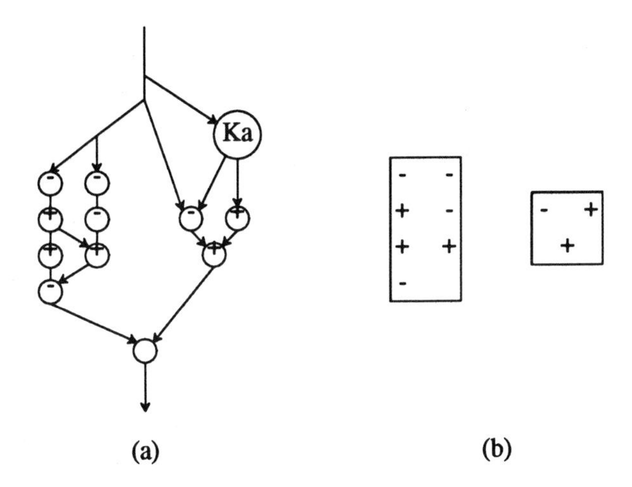

Figure 9.2 Partition of hardware for bounded unknown timing constraints. Three subtractors, three adders, and 6 control words are required.

Another approach possibly requiring a larger controller, would allow the interface independent code operations to have different schedules depending upon when the data from K_a arrives. This new problem requires OASIC to be solved two times. The first OASIC solution assumes that (data from K_a) arrives at the latest possible cstep. By using this schedule for interface independent code operations the second OASIC problem to solve involves scheduling and allocating hardware for conditional code (now representing both interface independent and dependent code operations) originating form each successive cstep within the interval p.

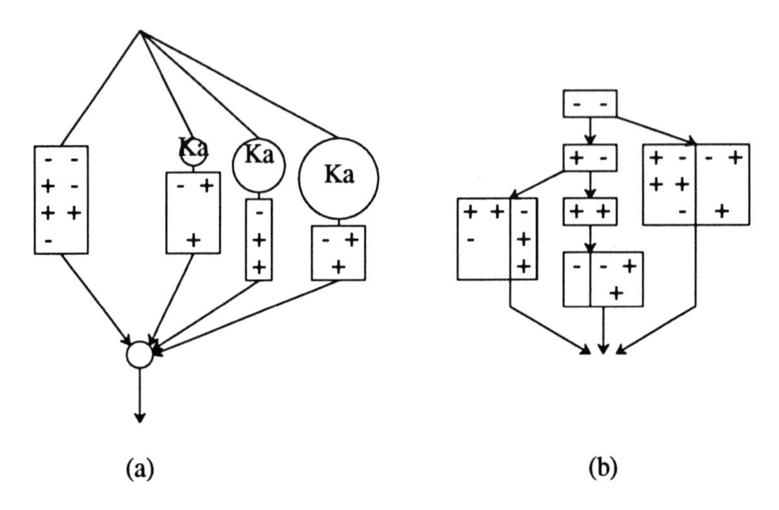

Figure 9.3. Sharing hardware between interface independent and dependent code operations. Two subtractors, two adders, and 11 control words are required.

d) Mutually Exclusive Pipelining

In the previous approach, c), we used p mutually exclusive branches (b=1,...,p) of interface dependent code operations where each branch starts at $d+b$ csteps after k_{out} outputs data $k_{out} <\bullet K_a$ (using fixed timing constraints for b=1). To minimize controller costs and without additional variables (unlike approach (c)), we can alternatively schedule the mutually exclusive branches as pipestages. This can be done by replacing $(\sum_{\substack{k \\ j \varepsilon R(k)}} x_{j,k})$ in (7.3),(7.4),(7.5) with the following equation:

$$\Big(\sum_{\substack{k \\ j \varepsilon R(k) \\ k\ indep\ K_a}} x_{k,j} + \sum_{\substack{k \\ K_a <\bullet k \\ (j-b) \varepsilon R(k)}} x_{j-b,k} \Big), \forall b$$

, where $k\ indep\ K_a$ represents code operations that execute in parallel and independent of the asynchronous process. This technique models the interface to asynchronous operations using p mutually exclusive pipestages, each with latency = 1 (see chapter 8 for terminology definition). An example is shown in figure 9.4.

To prevent the controller from becoming very large, as in 1(c), we have placed a restriction on the schedule of the interface dependent code. The restriction requires that the schedule for the interface dependent code must be the same for initiation at any time ε p. Thus only one schedule is required regardless of which time the data arrives and the algorithm is initiated.

We also could extend this approach by allowing interface independent code to have a different schedule depending upon when data from K_a arrives, as discussed in the previous section c). The same formulation is used, except the schedules for interface dependent code operations are constrained to be the same.

e) As Late as Possible Approach: Tradeoff time for area

In a) through d) we have assumed that as soon as the data from K_a arrives, the interface dependent algorithm is initiated immediately. However another approach is to wait for the data. The interface independent code operations are executed until the last cstep in interval p is reached. At this point the controller executes the interface dependent code operations along with the remaining interface independent code operations. This problem is very simple in OASIC and is modeled by placing a minimum timing constraint of $(C_{out}-1+d+p)$ between k_{out} and K_a.

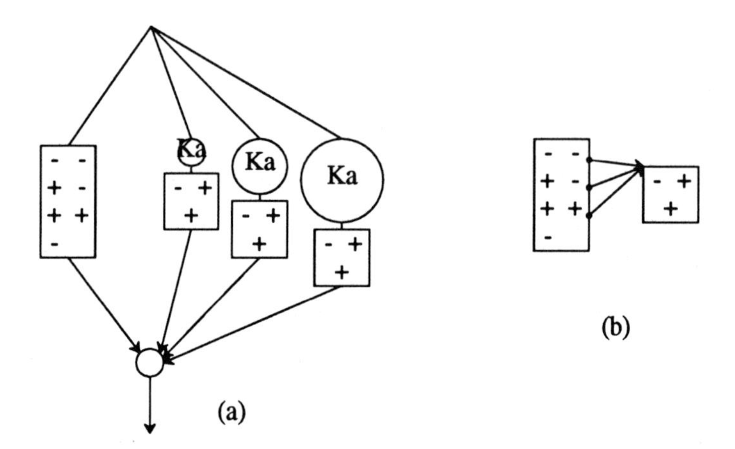

Figure 9.4. Asynchronous interface modeled as p mutually exclusive pipestages. Two subtractors, three adders, and 6 control words are required.

9.4 UNKNOWN UNBOUNDED DELAYS

Examples of unknown unbounded delays were presented in chapter 2. In some cases an ∞ bounded delay can be decomposed into a bounded delay and wait state. This partitioning was illustrated in figure 2.4 of chapter 2. In these cases we can use the previous models to determine which approach best matches the application. Otherwise when the partitioning is not possible it can be implemented as a wait state.

9.5 COMPLEX TIMING CONSTRAINTS

Timing constraints may also be formulated across loops. In other words operation a of loop iteration i is to be scheduled T csteps before operation b of loop iteration i+1. This example can be formulated as the following constraint (9.4), where k_{top} and k_{bot} are the top and bottom of the loop operations respectively as defined in chapter 8.2.

$$\sum_{j \varepsilon R(top)} (j+1)\, x_{j,top} - \sum_{j \varepsilon R(a)} j\, x_{j,a} + \sum_{j \varepsilon R(b)} j\, x_{j,b} - \sum_{j \varepsilon R(bot)} (j-1)\, x_{j,bot} = T. \tag{9.4}$$

The node packing formulation could not be made unless the loop execution time is known. In this case one can formulate the constraint as a fixed timing constraint using $J-T$ csteps for operation a to be scheduled after operation b of the same loop iteration.

Other combinations of constraints such as time(a,b) ≤ 0 or time(a,b) ≥ 3 can also easily be incorporated. These constraints cannot be modeled using a combination of minimum and maximum timing constraints and therefore the formulation in (Baker, 1974) and (Lee, 1989) cannot be used. Figure 9.5 illustrates one facet for this application, which can easily be generalized.

Other constraints such as ensuring that data is valid in a register for a least T csteps can be formulated using a dummy operation, k_d, and setting a minimum time constraint between it and the code operation ($<\bullet k_d$) which outputs the data. Alternatively if data is input from an external process and is only valid in an input register port for T csteps (after which point it may be overwritten), then the formulation is a maximum timing constraint between the dummy operation (representing the external process) and all operations ($\bullet > k_d$) which directly access this data.

In this chapter we outlined general and complex timing constraints that may be necessary for interfacing to external processes. We have described above for the first time, within the context of simultaneous scheduling and allocation, different approaches for dealing with

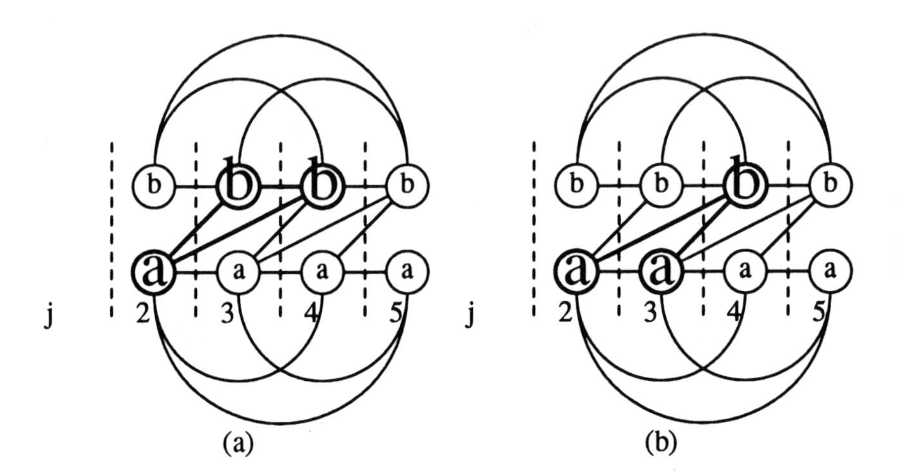

Figure 9.5. Node packing facets, in bold, for combination timing constraint *time*$(a,b)\leq 0$ *or time*$(a,b)\geq 3$.

interfaces to analog, asynchronous, and data-dependent processes. In the next chapter we will present applications and solutions of OASIC for various high level synthesis benchmarks and other algorithms, as well as demonstrating all types of interface constraints.

10.

OASIC SYNTHESIS RESULTS

This chapter presents results for simultaneous scheduling and allocation of a number of high-level synthesis benchmark examples (sections 10.1,10.3) (hlsw, 1988) , a digital neural network (perceptron with back propagation learning (Lippmann, 1987)) (section 10.2), and other examples to demonstrate interface constraints (section 10.4). The abbreviations for these examples are given in table 10.1 and more examples can be found in (Gebotys, 1991x) .

All the reported CPU times in chapter 10 are for solving the IP problems using GAMS/MINOS (LP solver) and/or GAMS/ZOOM (branch and bound IP solver) on a IBM PS/2 model 80 (386 PC). In examples where the GAMS/ZOOM branch and bound algorithm for IP is used, both the absolute and relative termination tolerances were set to zero (and upper bounds on variables were not set), so that in both cases globally optimal solutions could be guaranteed. The only times not reported are

Table 10.1. Summary of OASIC Synthesized Examples

Section		Example
10.1	EWF	Elliptical Wave Filter
10.2	ANN	Neural Network 16X4X4X4
10.3	CC	Conditional Code Example
10.4	IC	Examples with Analog and Asynchronous Interfaces

for preprocessing ie. translating the data flow graph or input algorithm into partial orders, calculating the asap and alap schedules, and reducing the edges for variable lifetime representation. The first translation of algorithm into partial orders can easily be done by traversal of the data flow graph or by writing a program to translate the input algorithm into a list of partial orders (Aho, 1974) . The asap and alap schedules algorithms were run below on the examples to give an indication of how quickly this processing can be done. The asap and alap schedules were run using the OASIC submodel ($x_{j,k}$), and some results are tabulated in table 10.2. We expect that by using graph theoretical algorthms such as the critical path method faster runtimes can be achieved. However it was more useful to demonstrate using the IP model to show that in fact all cases provided integer solutions, thus showing the tightness of the submodel. (It is also interesting to note that this is the first time an IP has been used to solve for these asap and alap schedules).

Table 10.3 summarizes the use of IP techniques. The use of these inequalities are application dependent and therefore are outlined below. The definitions and examples of their use can be found in chapter 6 and 7. The disjunctive constraints are only used with the structured OASIC model (chapter 6), whereas the tightened constraints and knapsack facets are only used with the area-delay OASIC model (chapter 7). The

Table 10.2. Preprocessing CPU seconds for EWF

Preprocessing	Te	CPU sec	
		Tgen	Texec
asap	21	186	46
alap	21	186	44

Table 10.3. Summary of Techniques Used in OASIC Examples

Technique	Sections/Examples			
	10.1 EWF	10.2 ANN	10.3 CC	10.4 IC
Tightening	y	n	n	y
Disjunctive	y	n	n	y
Knapsack	y	n	n	n
Functional Pipelining	n	y	n	n
Regularity	n	y	n	n

functional pipelining and regularity decomposition can be applied to either model.

The elliptical wave filter benchmark was thoroughly analyzed. Not only is it an excellent example for synthesizers; it has been a very popular benchmark for over three years, however OASIC results show that the architectures obtained by previous state of the art synthesizers are not globally optimal. In comparison to other benchmarks it has been synthesized much more often. Furthermore we have not seen any reported (published or unpublished) synthesized results for the matrix multiplication as performed in the kalman and artificial neural network examples.

10.1 ELLIPTICAL WAVE FILTER

The EWF was originally selected from the data flow graph representation (transformed from the z-diagram representation) in (Kung, 1985) as a high level synthesis benchmark in 1987. The DAG was then corrected by Dr.P.Paulin in (Paulin, 1987) . The majority of synthesizers have used a random topology, with the exception of SPAID and some others (Lee, 1989) which used a register transfer file architecture. A comparison with the register transfer file architectures is made at the end of the chapter.

The EWF is essentially a loop with 34 code operations and over 56 precedence constraints. It remains a challenge for current synthesizers due to its complex interconnections. The large number of precedence constraints provides a good benchmark for demonstrating the register allocation constraints. As we will show OASIC provides for the first time globally optimal synthesized architectures with improvements over previous research in number of busses and registers.

The number of registers allocated does not include the IN and OUT registers shown in (Kung, 1985) . It does not make sense to allocate these also since these are dedicated registers. The bus allocation includes only one input per multiplication operation since the multiplications obtain the other operand from the ROM or coefficient memory.

10.1.1 Structured Model

The synthesizers were compared by the types of subtasks they perform and the execution time in cpu minutes or seconds for the EWF. Table 10.4 has columns for scheduling (Sched), functional unit allocation (FU), and register allocation (Reg). A y (yes) means that the subtask is completely performed. A c (calculated) means that the number of resources is calculated exactly. A e (estimated) means that the resource is estimated using some heuristic or it is somehow considered during the algorithm. The execution time in cpu minutes/seconds in table 10.4 are for the scheduling phase only for HAL (Paulin, 1989) (using a Xerox 1108 Lisp machine) and the simulated annealing (S.A.) runtimes (using a Vax 11/8650) (Devadas, 1989) . The OASIC cpu seconds are for the GAMS/MINOS (Brooke, 1988) model generation time plus execution time (Tgen+Texec) for a IBM PS/2 Model 80 (386 PC). The asap / alap or other preprocessing times were not included but are negligible (less than 1 cpu minute). All integer solutions were obtained using the OASIC in over 80% of these cases (using different cost functions).

Table 10.4. Synthesizers Comparison for EWF

Example	Synthesizer	cpu (Tgen+Texec)	Subtasks Performed Sched	FU	Reg
EWF	S.A.	4min	y	y	c
	HAL	4-6min	y	y	e
	OASIC	53sec	y	y	-
	OASIC	90sec†	y	y	y

† using unstructured register allocation $\alpha x \leq 2R, \alpha(0,1,-1)$.

Table 10.5 and 10.6 give a more detailed examination of the performance of the OASIC for synthesizing the EWF as the upper bound on the number of control steps (csteps) is increased and as the EWF loop is unrolled. The model generation time(Tgen), execution time(Texec) (both in cpu seconds), number of variables(Var), number of constraints(Eqn) and number of simplex iterations(Itns) to solve the LP are given. The number of two cycle multipliers (*), two cycle pipelined multipliers (*pl), and one cycle adders (+) are given. In tables 10.5 and 10.6 the following cost function was minimized

$$\sum_i \sum_k \sum_{\substack{j \\ j \varepsilon R(k)}} (cost\_adder(i)\, x_{i,j,k} + cost\_multplr(i)\, x_{i,j,k}) \quad (10.1)$$

, where cost_adder(i)=(0,0,0,2,4,6) and cost_multplr(i)=(31,62,93,0,0,0).

All solutions in table 10.5 were completely integer after solving the LP once except row three (19 csteps). In this case the times reported in table 10.5 include enumerating by selecting variables by hand until a globally optimal all integer solution was obtained. We also tried to solve this example totally by extracting facets. Using additional facets (clique facets and lifted odd cycle facets) we could improve bounds by 0.8% from objective value =316 to 318.75, however it was faster to enumerate. We note that in section 10.1.2 we present a solution to this example in even faster runtimes using the unstructured model.

Table 10.5. Synthesis using IP Model for EWF

csteps	*pl	*	+	Tgen (sec)	Texe (sec)	Var	Eqn	Itns
17	2		3	10	9	173	113	115
18	1		3	15	11	267	157	193
19	1		2	40	29†	355	201	754†
17		3	3	10	10	184	127	111
18		2	2	14	12	267	163	172
19		2	2	14	20	355	207	263
21		1	2	37	20	537	287	629

† relaxed LP did not provide all $x \varepsilon B$, cpu times include B+B.

In table 10.6 we unrolled the EWF two and three times to illustrate how well OASIC performs with a large number of code operations. In the later case simultaneous scheduling and functional unit allocation of over 100 code operations was executed in 90 cpu seconds. To our knowledge no other research has solved simultaneous scheduling and allocation to global optimums for this (large) number of code operations.

Table 10.6. Synthesis using IP Model for Unrolled EWF

# of code operations	Te	*pl	+	Tgen (sec)	Texe (sec)	Var	Eqn
68	34	1	3	32	25	566	333
102	50	1	3	51	39	859	506

Figure 10.1 shows one solution for the elliptical wave filter example optimized for registers, functional units and execution time. This optimized solution was obtained by minimizing the previous area cost function (10.1), with an upper bound of 19 control steps, three adders, three two-cycle multipliers, and nine registers. The optimum solution with 2 two-cycle multipliers, 2 adders, and 9 registers (not including the IN and OUT registers) required 200 cpu seconds for model generation and 18 cpu seconds for LP execution (424 variables, 279 constraints, 536 iterations). Lifetime defining edges for all but two variables were found using the transitivity and alap analysis. The multiple edges for the two variables required only 24 extra constraints. No previous research to our knowledge have quoted as low as 9 registers for the EWF which demonstrates that global optimums have not been obtained by heuristic synthesizers. Other synthesized results with a constraint on the number of registers is shown in table 10.7.

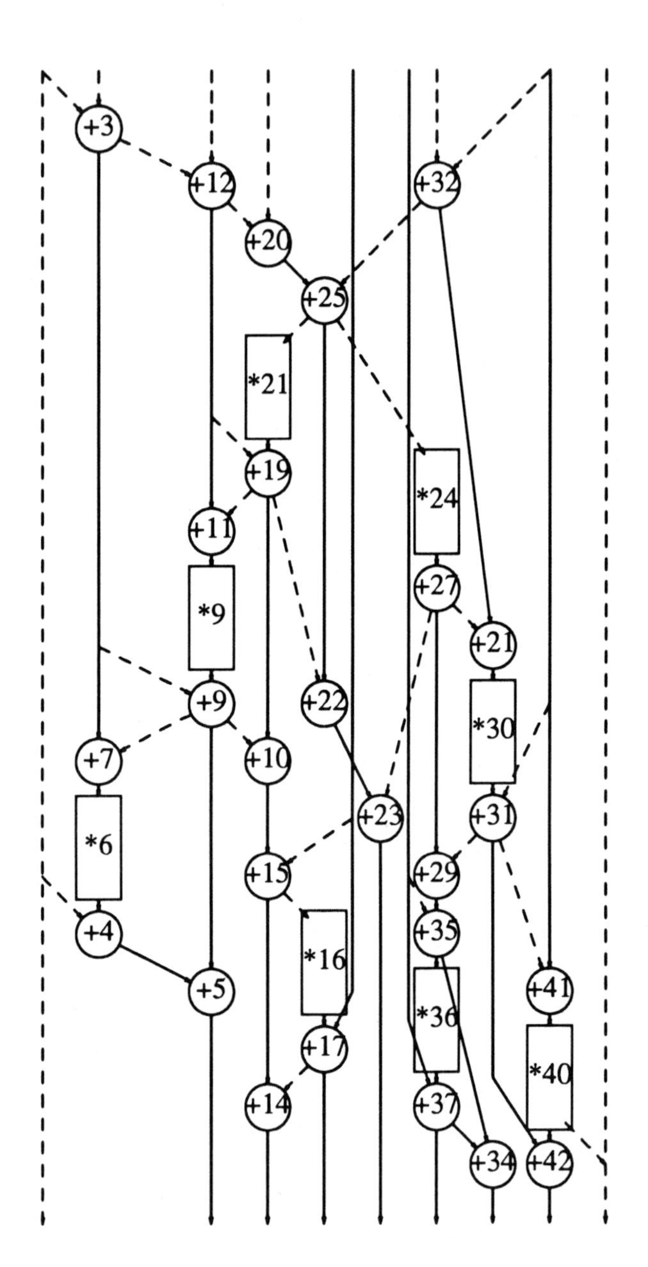

Figure 10.1. EWF schedule optimized for J=19 control steps, with variable lifetime defining edges.

Table 10.7. Structured Model with Register Allocation, $\alpha x \leq 2R$.

Te	*pl	*	+	R	Var	Eqn	cpu† (sec)
18	1		3	10	344	211	91
19		2	2	9	424	279	219

† Tgen+Texec cpu times

Table 10.8. Selection of Type of Functional Units

Cost per Type *	*pl	*	*pl	+	Tgen (sec)	Texec (sec)
12,25	31,62	2		2	16	13
19,37	31,62	1	1	2	16	16
Disjunctive Constraints						
150	250	2		2	18	(B+B)884
150,160	250,260	2		2	19	(B+B)475

Selection of Functional Unit Types.

To demonstrate the usefulness of the $x_{i,j,k}$ model we solved the OASIC model for simultaneous selection and allocation of functional units and scheduling. The results are given in table 10.8. Special cost functions calculated as described in chapter 6.3 (row 1,2) and disjunctive constraints (row 3,4) were used to select the type of functional units to minimize the area cost. In all cases the upper bound on csteps was 18 and the branch and bound algorithm was used to solve for the disjunctive variables. In the last row the cost values were incremented by 10 for each additional unit. This improved the cpu times and removed the ambiguity between choosing the first or second functional unit of the same type each with the same cost.

The selection of functional unit types is complicated by the fact that the lower bounds on the types of the functional units must be 0, (thus allowing for the case where the particular type is not chosen). Also depending upon the cost function more than one type may be selected. For the first time we have a model which can simultaneously make these decisions.

10.1.2 Area-Delay Optimized

No other synthesizer can simultaneously schedule and allocate all resources with an area-delay cost function except for OASIC and the simulated annealing synthesizer in (Devadas, 1989) . Bearing this in mind we compared OASIC with simulated annealing (Devadas, 1989) and the only two completely published solutions in HAL (Paulin, 1989) , and SAW (Lagnese, 1989) . We minimized area-delay cost functions with different upper bounds on csteps. The improvement in the number of busses for OASIC was compared with other published solutions. We also show that these solutions are relatively stable with respect to large changes in cost parameters.

Functional unit and bus allocation constraints were tightened where possible with k_{sep} = +25, and k_{out} of the EWF DAG shown in figure 10.1, using the technique described in chapter 7.6. Two solution techniques, LB and KP, were demonstrated with OASIC. The first method, LB, calculates lower bounds and branch and bounds to obtain a solution. The second method, KP, additionally uses knapsack inequalities to improve the bounds before branch and bounding.

Lower bounds on the number of busses and previous research are plotted in figure 10.2. Lower bounds are calculated by integer rounding up the minimum value of B obtained from solving the relaxed LP. The solid line in figure 10.2 is the lower bound (LB) for busses (without using $3I_+ \leq B$), the dashed line is the improved lower bound (LB+) using constraint $3I_+ \leq B$, and the circle points are the solutions obtained by OASIC. Lower bounds for busses calculated by fixing the number of functional units for each cstep (17,18,19,20,21) provided lower bounds of (10,8,6,6,6) number of busses respectively (Gebotys, 1991c, Gebotys, 1991b) .

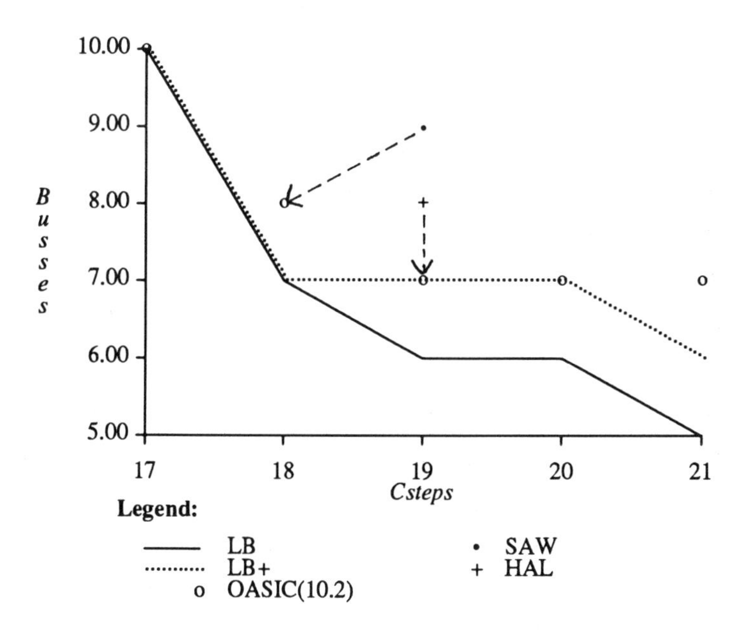

Figure 10.2. The graph of csteps versus the number of busses. The dashed arrow shows the improvement in busses using the OASIC model for designs with the same number and type of functional units.

The piecewise linear area-delay cost function (10.2), was minimized to synthesize architectures for the EWF, shown in rows OASIC of table 10.9. These cost parameters were taken from (Devadas, 1989) . The values shown in (10.2) were actually divided by 100 to try to normalize the objective function and hence improve the performance of the LP (Gill, 1981) . For example the optimized solution for 18 csteps with this area-delay cost function is 1 two-cycle pipelined multiplier (*pl), 3 (single cycle) adders, 9 busses, and 10 registers.

$$50\, I_{+} + 250\, I_{*} + 15\, R + 100\, B + 50\, T_e \qquad (10.2)$$

Figure 10.3 shows 7-18% improvement in area-delays over previous state of the art EWF solutions, SAW (Lagnese, 1989) and HAL (Paulin, 1989) for cost function (10.2). These results are very good considering the EWF has been investigated for years. Secondly our solution is stable over different types of cost parameters on the number of registers and busses.

Table 10.9 shows a comparison of the cpu performance using OASICs LB and KP methods with previous research. The - in table 10.9 means that the KP approach could not improve the lower bound on the busses. The delay cost component was removed once the first LP was solved with lower bounds. The largest cstep with code operations (or fractional values of these) was used as the delay value. Then subsequent analysis was done with minimizing area cost functions.

The only complete published solutions for the EWF were found in HAL (Paulin, 1989) and SAW (Lagnese, 1989) . From the 17 cycle schedule given in (Devadas, 1989) , although not specified, the eighth row requires 11 busses and OASIC requires 10 busses. The 17 and 18 csteps OASIC solutions for pipelined multipliers given in table 10.9 required 0.5 cpu minutes and 3 cpu minutes respectively where after branch and bounding on I_i,R,B variables, the initial LP provided all $x_{j,k}$

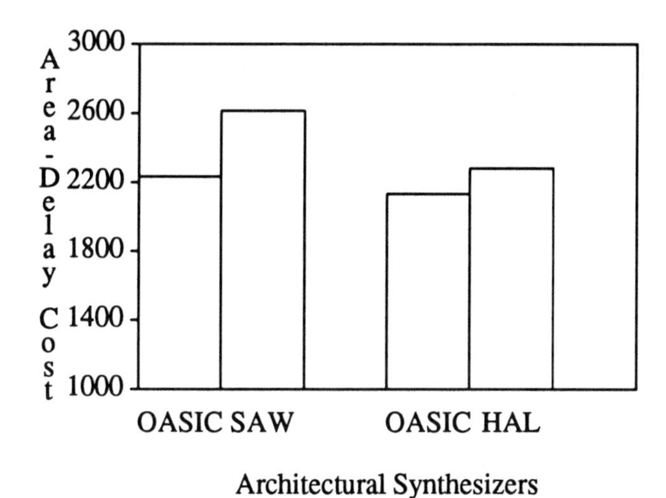

Figure 10.3. Comparison of synthesized solutions for EWF measured by area-delay cost(10.2) for architectural synthesizers OASIC, SAW and HAL.

integer solutions. These cpu times are faster than the 2 cpu minutes and 4 cpu minutes respectively quoted by HAL (Paulin, 1989) and simulated annealing (Devadas, 1989) .

HALs EWF synthesis for 19 csteps (Paulin, 1989) requires 8 busses and 12 registers (in 6 cpu min), unlike the optimal OASIC with 7 busses and 9 registers. This architecture was synthesized in less than 6 cpu minutes (including lower bounds calculation and branch and bound cpu times) to produce the schedule shown in figure 10.4. The lower bounds calculated were exact (7=B,9=R).

The area-delay optimized solution for two cycled multipliers (with an upper bound of 18 csteps) is shown in row seven of table 10.9 and plotted in figure 10.2. This is an interesting problem since it demonstrates the differences in lower bound calculation (LB), the advantages of the knapsack inequalities (KP) and importance of tight bounds. We will

Table 10.9. EWF Synthesized Architecture Comparisons

Synthesizer	Te	*pl	*	+	R	B	Total cpu minutes		
								LB	KP
OASIC	**21**		**1**	**2**	**9**	**7**		**30**	**6**
SAW‡	19		2	2	11	9	na		
HAL†	19	1		2	12	8	6		
OASIC	**19**	**1**		**2**	**9**	**7**		**5.8**	**-**
HAL	18	1		3	12	na	4		
OASIC	**18**	**1**		**3**	**10**	**9**		**3**	**-**
OASIC	**18**		**2**	**2**	**10**	**8**		**3**	**0.5**
HAL	17	2		3	12	na	2		
OASIC	**17**	**2**		**3**	**10**	**10**		**0.5**	**-**
OASIC	**17**		**3**	**3**	**10**	**10**		**0.5**	

† 6 busses + 2 local busses(Paulin,1989);‡ page 79 in (Lagnese,1989) na=not available;R does not include IN and OUT registers of filter.

now discuss in further detail the results of the different approaches.

By calculating lower bounds (2,2,9,7) for 18 csteps and branch and bounding on I_i,R,B variables, we obtained globally optimal schedule and allocation for this cost function in less than 3 cpu minutes total. In this case final relaxed LP with integer I_i,R,B variables also provided integer values for all $x_{j,k}$. In the LB method (row one of table 10.10) we required constraint, $3*I_+ \leq B$, to cut off the integer infeasible solution (I_+=3,I_*=2,R=9,B=7). The cpu times for bound calculations were approximately 10 cpu sec (Texec) for each variable in both methods.

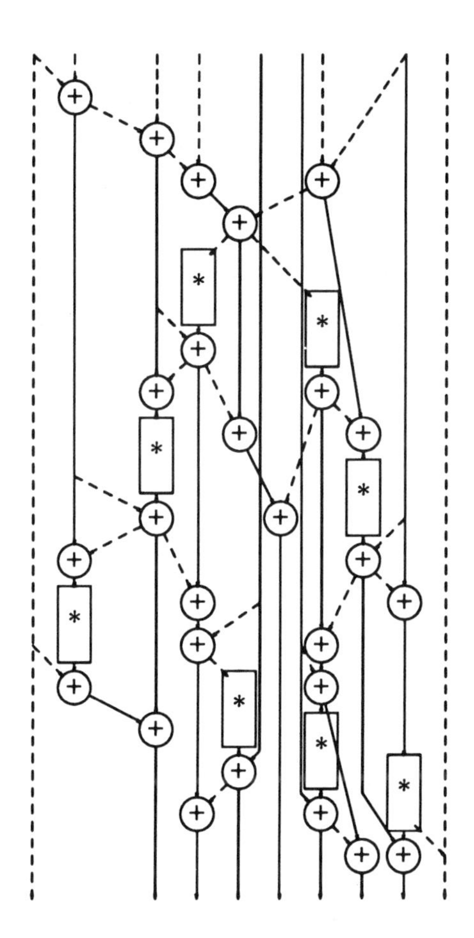

Figure 10.4. EWF schedule optimized for J=19 csteps and shown with variable lifetime defining edges. The rest of the data transfers are shown with dashed lines. Two adders, one two-cycle pipelined multiplier, 9 registers and 7 busses were allocated.

Knapsack Inequalities.

Since the lower bounds for 18 csteps (with two cycled multipliers) originally calculated were not exact (or equal to the final optimal values) we were able to use knapsack inequalities to improve the bound on busses. By fixing the lower bound of busses at 7, we were able to extract knapsack facets of this constraint and use it to show (by solving the relaxed LP) that the bound is not feasible. The bus allocation constraint with 7 busses is:

$$3\sum_{+} x_{j,+} + \sum_{*} x_{j,*} + \sum_{*} x_{(j-1),*} \leq 7, \forall j.$$

Consider the bus allocation constraint for j=15. Let $x_{15,k}=y_k$, and $x_{14,k}=y_{\bar{k}}$. Let the coefficients of y be c_k where k = + or k = *. Consider the following minimal dependent set, $C \varepsilon$ {+1, +2, *1, *2}, $(\sum_{k \varepsilon C} x_C \leq |C|-1)$.

$$y_{+1}+y_{+2}+y_{*1}+y_{\bar{*2}} \leq 3$$

where +1= +5 , +2= +35 , *1= *16, *2= *40 (+5, +35, *16, *40 are names of EWF code operations and are shown in figure 10.1). We can now prove that the following tighter inequality, where $k \varepsilon E(C)$ is a facet:

$$\sum_{\substack{+ \\ 15 \varepsilon R(+)}} y_{+}+y_{*1}+y_{\bar{*2}} \leq 3$$

First we prove that: (1) $C \setminus \{j_1, j_2\} \cup \{1\}$ is independent, ie.

$$c_{+}+c_{*1}+c_{*2}=3+1+1=5 \leq 7$$

Secondly we must prove (2) $C \setminus \{j_1\} \cup \{p\} | p{:}\min\ j\ \varepsilon N \setminus E(C)$ is independent, ie.

$$c_{+2}+c_{*1}+c_{*2}+c_{*p}=3+1+1+1=6\leq 7$$

We can generalize this inequality to choose all possible knapsack facets for the bus allocation constraint using:

$$\sum_{\substack{+ \\ j\varepsilon R(+)}} x_{j,+}+x_{j,*}+x_{(j-1),*^{-1}} \leq 3,\ \forall *,*^{-1},j\varepsilon R(*),j-1\varepsilon R(*^{-1}).$$

The other minimal dependent sets are redundant. With these knapsack inequalities we can solve the new relaxed LP (in 9 cpu sec) to determine that the LP is infeasible. We therefore improved the bound to 8 busses and solved the branch and bound with new bounds to obtain an all integer solution in a total of 24 cpu sec. This shows the advantages of using knapsack facets in solving the IPs. The total time required including all stages, generation and execution times, was 0.5 cpu min.

The other knapsack inequalities were used for example with 21 csteps. The initial lower bounds were $I_{+}=2$, $I_{*}=1$, R=9, B=6. By using generalized knapsack inequalities extracted from the bus allocation constraint, we proved the bound to be infeasible. So by increasing the lower bound on busses to B=7, we could branch and bound to a completely integer solution in 6 minutes total. Without using the knapsack inequalities we required 30 minutes of branch and bound to find the same globally optimal solution. Both solutions branch and bounded on the pipelined multiplier operations to obtain an all integer solution.

Variable Selection for Branch and Bound.

For the upper bound of 21 csteps an arbitrary branch and bound (where the multipliers are not chosen first) required 38 cpu minutes (with knapsack improved bounds) to produce all integer solutions. However by branch and bounding only on the variable for multiplication operations an all integer solution in 351 cpu seconds.

Lower Bounds Calculation.

Table 10.10 illustrates the performance of using different techniques to calculate the bounds of the problem. We also tried fixing the number of functional units to calculate the bounds on the registers and busses. Clearly this approach for the EWF application produced exact bounds and overall good execution times. However this is not always guaranteed to produce globally optimal solutions. For example in some cases it may be possible to increase the number of functional units in order to decrease the number of registers or busses. Figure 10.5 illustrates an example where by increasing the number of functional units we can decrease the number of busses. In 10.5 a) two adders, two (two cycle pipelined) multipliers, and 7 busses are allocated. However at the expense of an extra multiplier we can decrease the number of busses by one. In this case b) would be the optimal solution if the cost of one multiplier was less than the cost of one bus. For piecewise linear cost functions this decision will be more complicated (ie. will the cost of the bth bus exceed the cost of the ith multiplier?) and therefore in general we cannot fix the functional unit lower bounds when calculating lower bounds for other resources.

Comparison with Baker's Model.

Table 10.11 shows the total cpu seconds required by the OASIC model for optimizing the area-delay cost function in (10.2) for functional units alone. In row one the relaxed LP with (integer rounded) lower bounds produced integer solutions in 36 cpu seconds (for model generation and LP execution). We also ran this same instance of the EWF problem using the precedence constraint (6.5*) (Baker, 1974, Lee, 1989) from chapter 6 which required branch and bound to find an integer solution in approximately 10 cpu minutes (BAKER) on the same 386 PC using the same GAMS solvers. In both cases solutions are globally optimal for this cost function. Row three illustrates how efficiently we

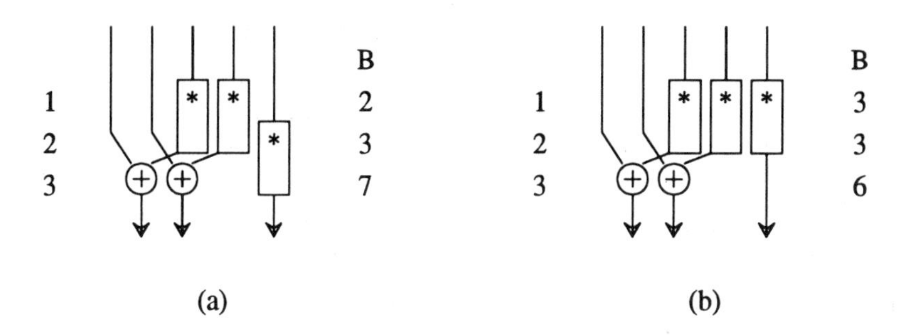

Figure 10.5. An example to illustrate the increase of functional units to decrease the number of busses. By increasing the number of multipliers by 1 in b) we can decrease busses from 7 in a) to 6 in b). The numbers on the left hand side and right hand side of each a) and b) indicate the csteps and the number of parallel data transfers respectively.

can simultaneously schedule and allocate large algorithms such as the EWF which was unrolled three times creating 102 input code operations. Over 300 $x_{j,k}$ variables were solved to integer values in the initial LP. These results illustrate how important good bounds and tight models are for solving integer programing problems.

Cost Function Sensitivity.

Changes in the cost parameters were also investigated to see how the cpu performance of the OASIC model varied. Furthermore this experiment provided more information on the stability of the allocations. This is important to determine since the area cost parameters are only an estimate of the final area values and if a great variation in allocations is produced by only a small change in the cost parameter then the designer must consider more than one allocation and schedule in the next binding phase and remaining design cycle. As shown in table 10.12, the

Table 10.10. IP Performance Comparisons of LB, Fixed, and Knapsack Met

Method		Lower Bounds				cpu
		I_*	I_+	R	B	
LB Method	LB	2	2	8.6	6.7	
	B+B	2	2	9	7	128
FU fixed Method	LB	2(fixed)	2(fixed)	9.3	7.3	
	rLP	2(fixed)	2(fixed)	10	8	38
FU fixed Method	LB	2(fixed)	2(fixed)	9.3	7.3	
	B+B	2	2	10	8	83
KnapSack Ineq	B+B	2	2	9	8(9.3sec)	24

Application: EWF with an upper bound of 18 csteps and 2 cycle multipliers.

Table 10.11. Comparison of cpu Seconds for EWF

Synth	# Code Operations	Te	*pl	+	Var	Eqns	cpu sec
OASIC	34	19	1	2	130	160	36
BAKER	34	19	1	2	130	120	600†
OASIC	102	50	1	3	310	407	40

† branch and bound is required.

solutions are very stable for the range of cost parameters.

Further Notes

1. The bounds calculated by fixing the number of functional units is exact and can save significant cpu time. However these do not guarantee that the solution is globally optimal since tradeoffs in registers or busses for functional units is not possible.

Table 10.12. Sensitivity Analysis for Cost Parameters

csteps	cost coefficients of				cpu seconds
	I_*	I_+	R	B	
18	250	50	15	100	24
18	250	100	15	300	26
18	250	100	300	100	152

2. A heuristic for 21 cycle EWF which assigned 7 partial orders between multipliers was used to solve the problem in even faster cpu times 14+78 cpu sec (execution + generation). It is possible that a branch and bound on partial orders would be even more efficient for these types of problems.

Comparison with Register File Architectures.

We could additionally make an estimated comparison with SPAID and other register file architectures. However as discussed in chapter 2 this is not a fair comparison due to the difficulty in judging the overall areas. For example we can compare the number of registers in OASIC with the number of registers in the register files, however the later requires less area per register. In addition we can compare the number of busses in OASIC with the number of busses in SPAID (Haroun, 1989) , however SPAID additionally requires multiplexors which OASIC does not. It is additionally difficult since the SPAID and <ESC> (Stok, 1989) compilers do not quote the number of multiplexors. Nevertheless from the number of inputs to multiplexors (mi) we can calculate a lower bound on the number of multiplexors m = (mi/RF) and total number of busses (B + m) = (RF + mi/RF). Since OASIC schedules and allocates, ie does not bind, we do not have a number of inputs to multiplexors to compare with. Table 10.13 compares the OASIC busses with the SPAID and

<ESC> busses plus multiplexors and the OASIC registers with the SPAID registers in the register files (Haroun, 1989) .

Table 10.13. Rough Comparison with Register File Architectures.

Synth	Te	*pl	+	RF	mi	B+m	R
OASIC	19	1	2	0	0	7	9
SPAID	19	1	2	5	17mi	9†	21
OASIC	17	2	3	0	0	10	10
SPAID	17	2	3	6	26mi	11†	21
<ESC>	17	-	-	8	23mi	11†	-

† OASIC lower bound on m was used in the calculation.

10.2 NEURAL NETWORK ALGORITHM

We will present results of the OASIC synthesizer for a four layer perceptron with back propagation learning (Lippmann, 1987) . It is an important example for synthesis since it is very different from the EWF in that it contains a great deal of regularity and has large loops. By using the regularity in the algorithm and by using functional pipelining constructs we can drastically reduce the number of code operations required to optimally schedule and allocate the filter. It is important to note that we are in fact not performing functional pipelining but only using the mathematical (functional pipelining) construction to reduce the number of code operations. The feedforward network is described first and we can show that the model is valid for all different sizes of neurons and layers. The second part of the results presents a model for back propagation. This is a more complex algorithm and analysis will vary depending upon the number of neurons in each layer.

The general algorithm for a four layer perceptron with back propagation learning is given in figure 10.6. If we let the input data width to be D bits ($D \geq 4$) then a D X 4 X 4 X 4 network is used, (one input, hidden, and output layer, where each layer has four neurons). This size was chosen so that we could demonstrate what the schedule and allocation would look like. Nevertheless we can increase the number of neurons per layer as described in the later part of this section.

main loop {

for each layer (input to output) {

for each neuron in a layer{

$$x_j^1 = f(\sum_i x_i w_{i,j} - \theta_j) \;\}\}$$

for the output layer{

$$\delta_k = x_k^1 \, (1 - x_k^1) \; (\partial_k - x_k^1)$$

for (j=1,...,4) $w_{j,k}^{(t+1)} = w_{j,k} + \eta \; \delta_k \; x_j^1; \;\}$

for the remaining layers (output-1 to input) {

$$\delta_j = x_j^1 \, (1 - x_j^1) \; (\sum_k \delta_k \, w_{j,k}^{(t+1)})$$

for (i=1,...,4) $w_{i,j}^{(t+1)} = w_{i,j} + \eta \; \delta_j \; x_i^1; \;\}\}$

Figure 10.6. Algorithmic description of ANN translated from (Lippmann, 1987) .

In order to take advantage of the regularity present in the NN algorithm we extracted a *stream* shown below in figure 10.7. In figure 10.7 x_d is the input data to the network, (f) and the next two lines are the forward propagation, and (1) through (3) and the next three lines represent the backwards propagation. This stream (or column of activity) in effect illustrates the behavior of the first neuron of each layer. The input code describing this behavior is given below. Equation (f) describes the feed-forward network that will be synthesized first. The remaining equations (1) through (3) describe the back propagation learning.

In order to avoid trivial analysis we examine a case with an even number of neurons per layer and we will allocate an odd number of multiplier and adder functional units. If we examine allocating an even number of functional units, F, and we have an even number of neurons per layer, N, then we can : (1) for F<N, execute N/F neurons in parallel using multiplier accumulator streams; (2) F≥N, execute F wide multiplier accumulator trees. An example of a multiplier accumulator stream is shown in figure 2.2a) of chapter 2. A three wide multiplier accumulator tree is also shown in figure 2.2b) of chapter 2.

Assuming we have 4 neurons per layer, we will examine the allocation of 3 functional units of each type (adders or multipliers). The initial multiplier accumulation is equivalent to matrix multiplication for a D X 4 matrix. Let us assume for simplicity that D=16. In the matrix multiplication, the summation loop was unrolled 16 times and the 4 columns were pipelined with a latency of 1. The 4 pipestages of 32 code operations (multiplier accumulator streams) were scheduled (using the functional pipelining constraints) in 24 control steps to execute the 16X4 multiplication, shown in figure 10.8. Three (two cycle) pipelined multipliers and three adders were allocated which can complete one pass in 215 control steps (requiring less than 2 cpu minutes to optimize). By making use of regularity, 4 pipestages of multiplier/adder streams with while($\exists x_d$) {

$$x_i^1 = f(\sum_d x_d \, w_{d,i} - \theta_i) \quad \text{(f)}$$

$$x_j^1 = f(\sum_i x_i^1 \, w_{i,j} - \theta_j)$$

$$x_k^1 = f(\sum_j x_j^1 \, w_{j,k} - \theta_k)$$

$$\delta_k = x_k^1 \, (1 - x_k^1) \, (\partial_k - x_k^1) \quad (1)$$

$$\text{for (j=1,...,4)} \quad w_{j,k}^{(t+1)} = w_{j,k} + \eta \, \delta_k \, x_j^1; \quad (2)$$

$$\delta_j = x_j^1 \, (1 - x_j^1) \, (\sum_k \delta_k \, w_{j,k}^{(t+1)}) \quad (3)$$

$$\text{for (i=1,...,4)} \quad w_{i,j}^{(t+1)} = w_{i,j} + \eta \, \delta_j \, x_i^1;$$

$$\delta_i = x_i^1 \, (1 - x_i^1) \, (\sum_j \delta_j \, w_{i,j}^{(t+1)})$$

$$\text{for (d=1,..,4,..,D)} \quad w_{d,i}^{(t+1)} = w_{d,i} + \eta \, \delta_i \, x_d^1; \; \}$$

Figure 10.7. Stream of ANN code representing behavior of the first neurons of each layer.

latency of one, required 41 $x_{j,k}$ variables and 66 constraints, were used to model this matrix multiplication. Apart from the memory required for input vector and matrix storage, there are 7 local registers required and 18 busses. OASIC could simultaneously schedule and allocate (3 multipliers and 3 adders in 24 csteps), in less than one cpu minute total. This included 14 cpu sec and 5 cpu sec for the model generation and the LP execution times. Unfortunately we cannot compare with (Lagnese, 1989) which performed matrix multiplication in a different application since they chain the multiplier and adder into one functional unit. No other published research has tackled matrix multiplication. This is most likely because of its size, complexity and the possible difficulties with applying heuristic synthesis techniques to an example with a great deal of regularity. Nevertheless the kalman filter example illustrates the flexibility of the IP model to support functional pipelining and synthesize different types of input algorithms. If each f(.)=table look up (since f(.) represents a nonlinear function) has the same number of csteps to produce an output then an upper bound of three f(.)s is also needed. The design exploration for this feedforward network is shown in table 10.14 requiring 20 cpu seconds to minimize an area cost function (10.2). We will present in this section synthesized architectures for the complete algorithm shown above in figure 10.8.

Back propagation involves performing equations (1) through (3) of figure 10.7, where (2) and (3) are repeated in a loop. It is not straightforward how one can schedule these equations due to their interdependence. A significant amount of parallelism is lost by separately synthesizing each equation separately. Furthermore unlike the feedforward results it is not obvious how one can extend this analysis to a 4X6 layered network (k=1,...6 and j=1,...,4). For these reasons we will demonstrate our synthesis model on a 4X6 network. Equations (2) and (3) represent 4 weight updates for each of 6 neurons and a mutltiplier accumulator stream of

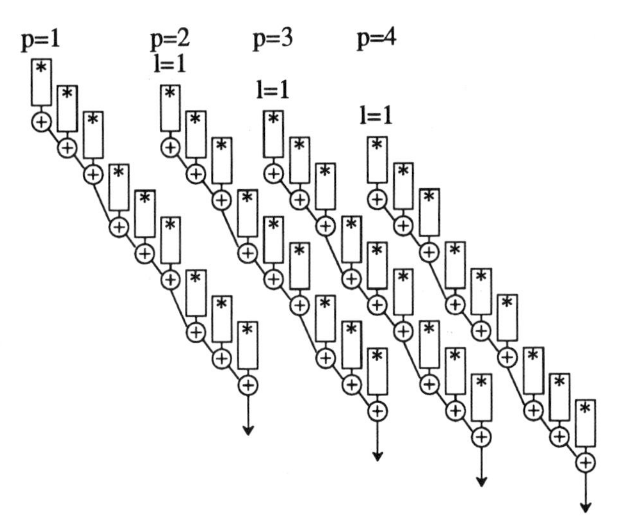

Figure 10.8. Part of the IP optimized schedule (shown for 4 control steps with 4 pipestages, latency=1, 7=R, 18=B) for the 16X4 matrix multiplication of the kalman filter benchmark example.

Table 10.14. Forward Propagation OASIC Results.

$x_i^1 = f(\sum_d x_d w_{d,i} - \theta_i)$						
Te	*	+	R	B	f(.)	-
24	3	3	7	18	-	-
26	3	3	7	18	3	3
28	3	3	7	18	1	1

length 6 for each of 4 neurons. The 4 weight updates over 6 neurons were transformed into 6 weight updates over 4 neurons. In other words each of the first 4 neurons (PE, k=1,..,4) had two additional weights to update (representing part of the 5th and 6th neuron weight updating). This provided the model with four (neurons) symmetric streams of code to functionally pipeline. These were then functionally pipelined to represent behavior of all neurons in one layer. The results are given in table 10.15, and the schedule of weight calculation is presented in figure 10.9. Each row of in figure 10.9 represents the accumulative summation over all k of $w_{j,k}$ performed by PE_j to calculate δ_j, output at the 6th cstep.

Table 10.15. Back Propagation OASIC Results for DX4X6 network.

for (j=1,...,4) $w_{j,k}^{(t+1)} = w_{j,k} + \eta\, \delta_k\, x_j^1;$ (2)

$$\delta_j = x_j^1\, (1 - x_j^1)\ \ (\sum_{k=1}^{k=6} \delta_k\, w_{j,k}^{(t+1)}) \qquad (3)$$

Te	*	+	R	B	Var	Eqn	Tgen	Texec
							(cpu sec)	
12	5	6	12	29	62	108	24	6
17	3	3	7	18	182	243	92	10

Cstep to calculate $w_{j,PE}$ by PE				
j , PE	1	2	3	4
1	1 , 5	2 , 6	3	4
2	4	1 , 5	2 , 6	3
3	3	4	1 , 5	2 , 6
4	2 , 6	3	4	1 , 5

at cstep 5,6 $w_{j,k} | k$=5,6 are calculated by PEs shown.

Figure 10.9. Schedule of weight calculations in back propagation for a 4X6 network.

10.3 CONDITIONAL CODE EXAMPLE

The benchmark example for conditional code was originally presented in (Kurdahi, 1987, Park, 1986) . The example is shown in figure 10.10, and contains five nodes where branches are initiated, and 15 code operations. It is a good example for synthesizing conditional code since there are partially ordered and independent conditional branches. We assume that each edge in the figure 10.10 represents a data transfer and conditional statements take 0 delay.

In addition to minimizing the area of the architecture, we can formulate it to minimize the execution times of each path. Only one other synthesizer (Camposano, 1991) can do this type of scheduling however it requires operations to be chained. Nevertheless this is very important since different paths will take different times and each may have an equal probability of being executed. If probabilities are not equal we can accordingly weigh the final execution time on each branch.

There are three conditional blocks in this example, see figure 10.10. We set these up as k εC_c^b, b=1,2,3. Each block has c number of branches possible. For example +1,+2,-2,+5 εC_1^1 +1,+2,-2,-5 εC_2^1 +1,+2,+3,-6

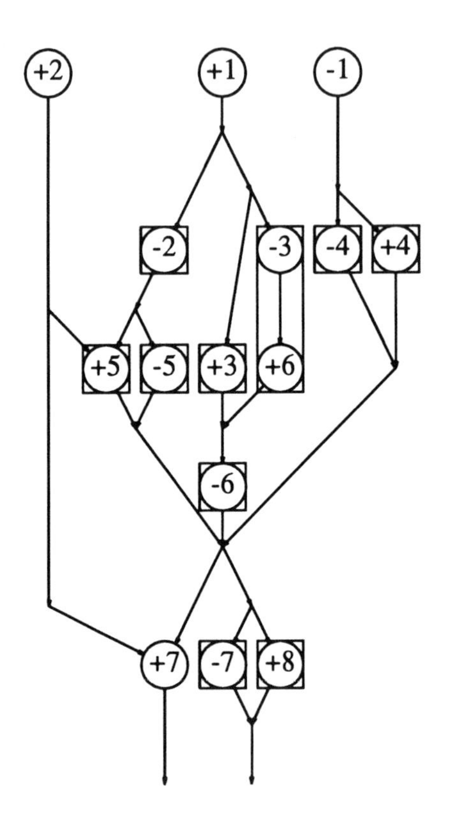

Figure 10.10. Conditional code example from (Kurdahi, 1987) showing a schedule for 2 adders, 2 subtractors, 3 registers, 9 busses, and Te=5.

εC_3^1 +1,+2,-3,+6,-6 εC_4^1 +4,-1 εC_5^2 -4,-1 εC_6^2 +7,-7 εC_7^3 and +7,+8 εC_8^3. Since b=1 and b=2 are in parallel we formulated constraints for all possible combinations of branches within these two blocks. However b=3 succeeds the two blocks so we generate only constraints for all branches in this one block separately.

Table 10.16. CPU Times for Minimizing Total Csteps in Conditional Paths

Te	+	-	R	B	CPU sec Tgen	Texec	Var	Eqn
5	2	2	-	-	13	10	90	302
6	1	2	-	-	13	10	90	302
6	1	1	-	-	13	10	90	302
5	2	2	3	9	145	16	92	680
6	1	2	3	9	145	16	92	680
6	1	1	3	6	143	16	92	680

$I_{+/-}$ were fixed and the execution times for all paths minimized.

Only one edge (originating from +2) was reduced by the algorithm since the variable lifetimes in different branches are calculated by the inequality. The remaining multiple edges in the DAG are part of a conditional branch. In the OASIC register allocation inequality when one branch is chosen only one lifetime defining edge is selected during constraint generation.

The allocation and schedule time for 1 adder, 2 subtractors, 6 csteps, 3 registers, and 9 busses are shown in table 10.16. In (Kurdahi, 1987) 10 csteps are used and 8 registers are used, however 8 inputs are used into the data flow graph. We omitted these inputs to show that our algorithm produces the optimal number of registers.

10.4 ANALOG AND ASYNCHRONOUS INTERFACE EXAMPLES

The use of OASIC with analog and asychronous interfaces is demonstrated in this section.

10.4.1 Analog Interface

To illustrate the use of the OASIC timing constraints to model an analog interface we used the successive inputs of IN in the EWF example. For the 2Xs unrolled filter we placed a timing constraints between "+3" (see figure 10.1) in successive loops to be scheduled exactly 17 csteps apart. The synthesized schedule and allocation with the fixed timing constraint required 60 cpu seconds (total generation + execution time) as shown in table 10.17, for the structured model (of chapter 6). The area cost function was minimized and the relaxed linear program produced all integer variables. One extra multiplier is required for this architecture due to the analog interface. The solution without fixed timing constraints was presented in table 10.17 of section 10.2.2.

Table 10.17. Example of Analog Interface for unrolled EWF.

Code Opns	Te	*pl	+	CPU sec	Var	Eqns
68	34	2	3	60	566	337

10.4.2 Asynchronous Interface

To demonstrate how the asynchronous interface can be solved we used the EWF example and replaced +25 <• *24 with +25 <• k_a and k_a <• *24. The asynchronous process k_a has a minimum data processing time of 2 csteps (d=2) and a maximum of 5 csteps before an output data value is produced (p=3, defined in chapter 9). We used an upper bound of 21 csteps for the new filter and minimized the area cost function using the OASIC model from chapter 7.

First the asynchronous interface was modeled using 3 mutually exclusive pipestages (see chapter 9, section 9.3 d)). Since the interface dependent code will be pipelined we only use variables to represent the first pipestage where k_a requires 2 csteps to produce output data.

Preprocessing for the interface dependent code is done as usual. The upper bound on csteps for the interface dependent code is J^{UB} - 4 (to account for the 2 cstep delay of k_a, and 2 csteps for the two other pipestages). In this example the interface dependent code does not precede other interface independent code, so J^{UB} is used as an upper bound for the interface independent code.

Table 10.18. Asynchronous Interface Example.

k_a	Method	Te		*	+	CPU	Var	Eqns
d,p		LB	UB			sec		
2,0	Fixed	21	21	1	2	25†	196	233
2,3	PL	19	21	2	2	38	140	230
2,3	CC	20	21	2	2	137	217	331
4,0	WC	21	21	2	2	20†	140	174
2,3	PL	25	27	1	2	1618	311	454
2,3	CC	23	23	1	2	45†	335	473
4,0	WC	23	23	1	2	28†	206	260

† Tgen+Texec of relaxed LP where $\forall$ x εB.(other CPU for B+B)

The same example was scheduled with mutually exclusive conditional code (CC, see chapter 9, section c)) thus allowing for possible improvements in hardware at the expense of controller area. The comparisons of pipelined (PL) and conditional (CC) solutions for this example are given in table 10.18 (see chapter 9 for a definition of d and p). In addition the solution without an asynchronous interface, but with a fixed timing constraint of 2 csteps (between +25 and *24) is given in the first row (Fixed). The fourth and last rows of table 10.18 show the worst case examples (WC), where k_a requires 4 csteps to produce output data. In the CC case the schedule of each stream was not identical as in PL. In one case by using the '+25' operation to tighten some inequalities the branch and bound required 13 nodes in the branch and bound tree and 267 cpu seconds to determine that the IP was integer infeasible. Without using the tightened inequalities the same problem required 22 nodes in the branch and bound tree and 420 cpu seconds. We have shown that both approaches for asynchronous interfacing are practical with respect to its solution in the OASIC model. The schedule for the PL solution is shown in figure 10.11.

OASIC allows the flexibility of analyzing different approaches to synthesizing architectures in the presence of complex interfaces. For example one possible methodology to follow may be to synthesize for minimum controller costs using the PL strategy and then calculate lower bounds with the CC strategy. If the lower bounds are equivalent to the PL schedule then there is no advantage to a larger controller. Alternatively one may wish to investigate any savings with the CC method.

In chapter 10 we have used OASIC to synthesize architectures and analyze a number of input algorithms or benchmarks. Globally optimal solutions which minimize an area or area-delay cost function have been synthesized in practical execution times. A summary and discussion of these results will be provided in the next chapter. Some concluding remarks and future extensions will also be discussed.

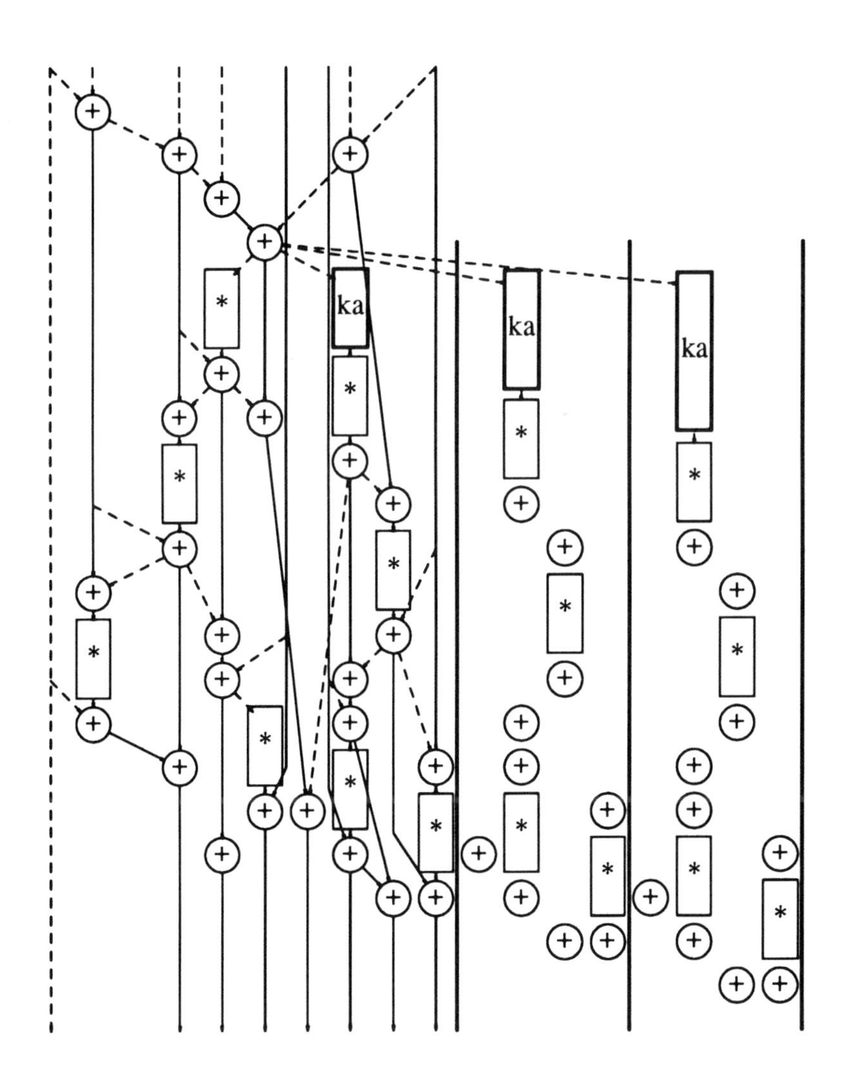

Figure 10.11. Optimized schedule and allocation for J=21 control steps, with asynchronous interface to k_a. Mutually exclusive schedules are shown separated by bold vertical lines.

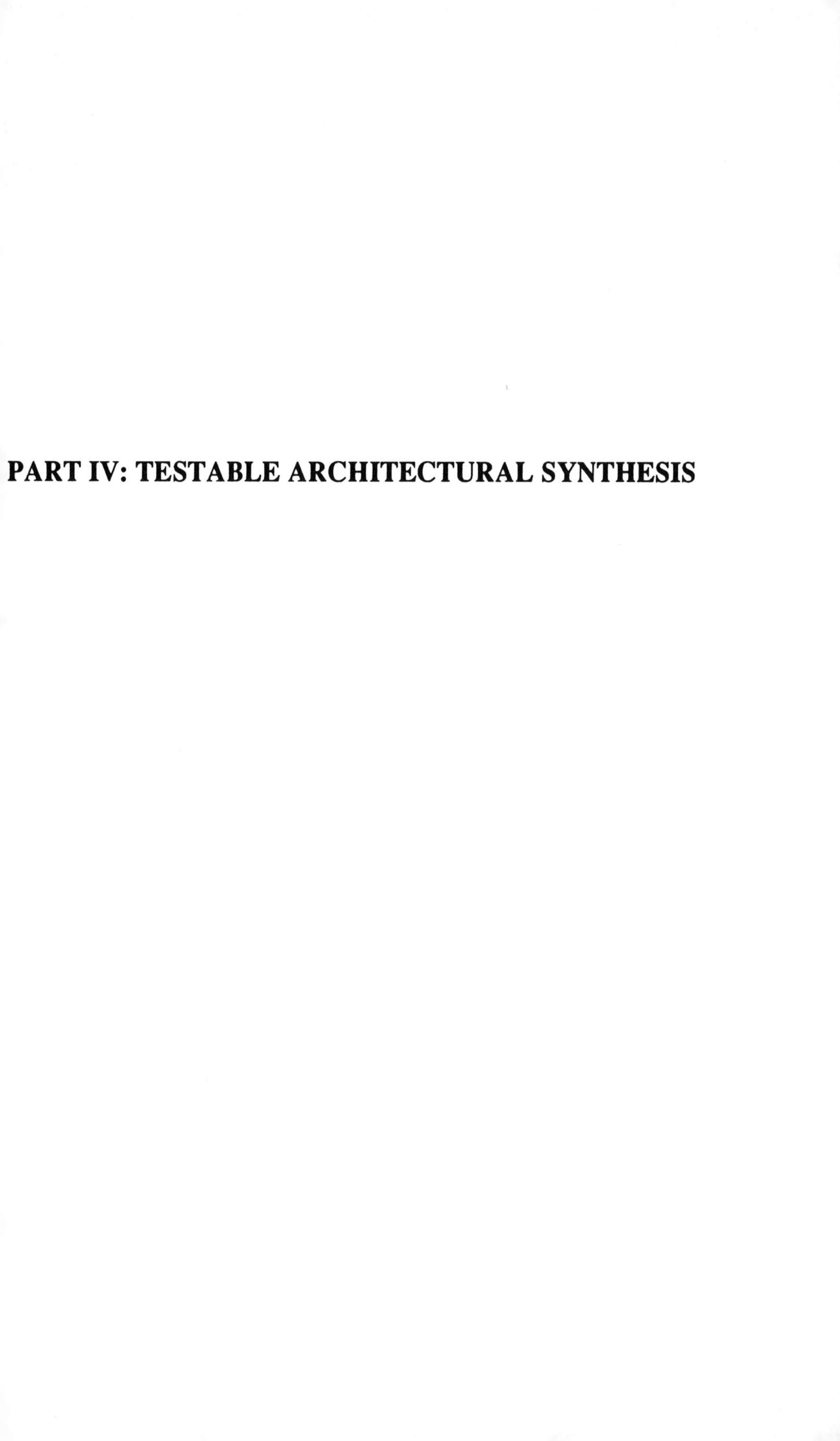

PART IV: TESTABLE ARCHITECTURAL SYNTHESIS

11.

TESTABILITY IN ARCHITECTURAL SYNTHESIS

The need for early test consideration during digital design is well recognized and documented by the VLSI Industry (Tsui, 1986) , (Williams, 1983) , (Fung, 1986) , (Abadir, 1985) . However there have only been a few recent approaches to integrating these two areas : in particular architectural design synthesis and test incorporation (Fung, 1986) , (Abadir, 1985) , (Gebotys, 1989) . In this chapter we will define, discuss, and compare problems in both areas. This chapter is not a tutorial on testing, however test references are given to aid the reader. Topics affecting the test incorporation for complex VLSI digital designs and test issues affecting the design synthesis methodology are covered. Advances in test research have a direct effect on the way in which testability can be implemented in a synthesized design. It also directly effects the testable design synthesis methodology, as discussed in chapter 1. Approaches to previous research in test incorporation, section 11.2 and

11.3, the integration of design and test, section 11.4, and a list of remaining problems in this area, section 11.5, will be covered in this chapter.

11.1 DESIGN AND TEST

The objective of integrating architectural synthesis and test incorporation is to provide an optimal architecture that is testable. Furthermore by integrating these two tasks, an architectural design solution will be found that satisfies both design and test constraints such as area, delay, testing time, and estimated fault coverage.

The integration of design and test is aimed at decreasing the VLSI design cycle time and avoiding large time consuming test efforts at the end of the design cycle. Test should be considered early in the design cycle during design and not after.

A comparison of the architectural synthesis problem and the structured design-for-test problem is given in table 11.1. The objectives, methods, difficulty, and levels of design representation for architectural synthesis and test incorporation are outlined. It is clear that some aspects of these problems are similar and overlap such as the problem constraints, and the problem tasks.

The constraints are quite similar except the overhead in area and speed (OH) of the architecture due to the incorporation of test structures is used as a measure in the second column of table 11.1. The testing time is defined as the number of clock periods required to test the architecture. The fault coverage is an estimate of the fault coverage of the architecture that is achieved by incorporating testability. For example if one allows a very long testing time, many test vectors can be applied and therefore the fault coverage will be increased.

Table 11.1. Comparison of Synthesis and Test at the Architectural Level

Problem	Architectural Synthesis	Test Incorporation
Constraints	Area, Speed, Execution Time	Area OH, Speed OH, Test Time Fault Coverage (Test Confidence)
Task	Schedule, Allocate Hardware, Binding.	Test Schedule, Allocate Test Hardware, Bind Test.
Verify	Functional Simulation	Fault Simulation
Design Level	Functional Units, Registers, Busses, Multiplexors.	Combinational Units, Sequential Units, Busses, Multiplexors.
Difficulty	NP hard	NP complete

The tasks for architectural synthesis have been already outlined in chapter 3 and part III of this text. Nevertheless it is very interesting to compare these with test incorporation tasks at the architectural level. In architectural synthesis code operations will be executed during a particular clock period. In test incorporation each hardware resource (bus, register, functional unit) will have a particular *test phase* during which it will be tested (or special vectors will be present at the inputs and processed at the outputs). In architectural synthesis functional units, registers, and busses are allocated, whereas in test incorporation test registers (which do not replace existing design registers) and special interconnect (ie. single bit width for serially shifting test data or full word widths for parallel transfer of test data off or on the chip) may be allocated. In the later case allocation of test interconnect may include general bus structures, such as the allocation of multiplexors, which are only used during the testing of the chip. Binding in architectural synthesis refers to the assignment of code operations to functional units and variables to busses and registers. For test incorporation, we use the term *test binding* (see table 11.1) to refer to the assignment of registers to test registers. In other words existing design registers allocated from architectural synthesis, are replaced with test registers. *Test registers* is the generic term we use to identify special hardware that during the normal operation mode may or may not act as registers (in the later case they are inactive), and during the test mode of the chip either produce or access test vectors for input to or output from functional units. These will be further defined in section 11.2. We use the term *test binding* to differentiate from the allocating of test registers. In the later case each test register allocated does not replace any existing design registers.

Verification can be performed through functional or fault simulation for architectural synthesis or test incorporation respectively. The design levels, shown in table 11.1, are similar since we are only interested in the architectural level of the design automation. Lower levels of testing and design are briefly discussed in chapter 1. The architectural synthesis problem and test incorporation problem are NP hard and NP complete respectively.

11.1.1 Choices in Design and Test

There are three choices for incorporating test into architectural synthesis.

1. No test consideration during the design.
2. Leave test consideration until after the design is synthesized.
3. Try a structured simultaneous approach to both architectural synthesis and test incorporation.

Choice 1 uses the functional testing to provide a test, with no modifications to the hardware. Choices 2 and 3 are shown in figures 11.1 and 11.2. Figure 11.1 shows five blocks: the architectural synthesis process is represented by the first two stages, followed by the test incorporation, and finally the design placement, routing, and layout stage. The first two stages could also be represented as one stage in a synthesizer such as OASIC (see part III) which minimizes a area-delay cost function to synthesize the architecture. Feedback in this case could be the selection of a different clock speed, and therefore different functional units or performing higher level input algorithm transformations to extract more parallelism for the synthesizer. The separation of the design and test stage has been proposed for systems, such as (Abadir, 1985) , where the architectural research (Granacki, 1985) and test research is performed by separate groups. Other researchers have also proposed this, such as (Beausang, 1987) . The test overhead in figure 11.1 refers to the additional area and delay of the design required by the test incorporation. This change in constraints is viewed as an overhead because the synthesis exploration is finished and the architectural solution before test is essentially fixed, due to lack of feedback after test. Figure 11.2 refers to further integrating architectural synthesis with test such that the design exploration is guided by test constraints in addition to the design synthesis constraints. In this methodology there are no overheads since there is feedback after test. Only constraints for a testable design such as area, delay, and test cost (which may include estimated fault coverage and test time) are evaluated. In all cases test incorporation refers to the analysis, or modification of the design synthesized for test. In figure 11.2 design and test are integrated together the presence of feedback after the test stage to the synthesis process as in (Gebotys, 1989) thus providing testable design exploration. Some systems (Fung, 1986) have integrated design and test in a finite state machine environment however no methods for feedback are presented. The layout box actually refers to the remaining design activities required after an architectural solution is formed. For example this would include placement, module generation, routing, and final layout or mask generation for fabrication. Again these tasks vary depending upon the technology used (ie. gate array, standard cell) and level of design output from the synthesizer (including floorplan or whether a netlist only is provided).

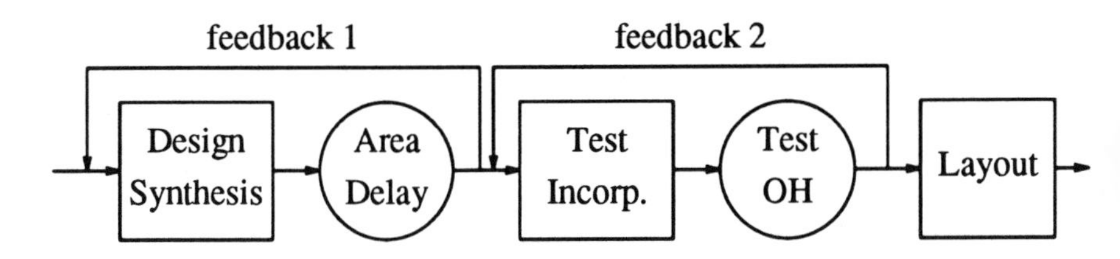

Figure 11.1. The design synthesis process and independent test incorporation process shown with constraints guiding feedback.

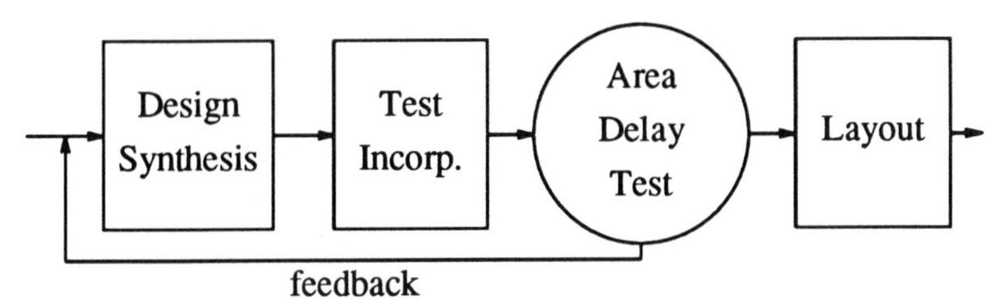

Figure 11.2. Integrated design synthesis and test incorporation. Feedback after test incorporation used for design exploration.

When no test is considered during design, in choice 1., a functional simulation is used to test the designs. To verify that the synthesized design works correctly, a functional test whose vectors represent all possible inputs would be created. For combinational chips with M inputs 2^M test vectors would be required for full functionality. However this may not detect sequential faults, for example a slow-to-rise fault, present in the circuit . To test for sequential faults in theory one would require all possible orderings of all test vectors which is combinatorially explosive. Furthermore the design may contain untestable constructs such as redundant circuitry because test was not considered during design.

For functional testing of sequential chips the test would have to include all different sequences of pattern inputs each of which may be replicated an undetermined number of times. This clearly may not be practical, especially for circuits with more than 25 inputs (Susskind, 1984) . A minimum of 2^{N+M} would be required, for M inputs and N latches in the design, but this set of vectors does not even modify the sequence of inputs, (Williams, 1983) . Furthermore in real applications where the chip is interfaced to unknown or complex external processes, the set of all functional tests possible may not always be obtainable. Thus a functional test is often not possible to define and is not sufficient to test the chip with confidence.

The second case, shown in figure 11.1, considers both architectural synthesis exploration and test exploration in a separate and unintegrated manner with no feedback between these two tasks. The main problem with this approach is that complicated designs may be impossible to test without changes to the hardware. These hardware changes may cause the design constraints to be exceeded or cause high test overheads. Thus a design solution found in the synthesis stage may no longer be valid or meet desired constraints after the test exploration stage. Thus with no feedback to the architectural synthesis the DA tool would fail to find a solution although one may exist.

The exception to this, is the case where the synthesized design solution before test incorporation is well within its design constraints. In other words during architectural synthesis the minimum area and delay of the final architecture were less than the designers requirements. For example the synthesized architecture may be modified for testability by adding scan registers and additional interconnections. The area and delay values for the new testable architecture are then computed and may still meet the original design specifications. There are two problems with this approach. The first problem is how to estimate the test overheads, especially if the designer does not know which test method is to be applied. These overheads will vary depending upon the design solution and the test methodology to be implemented (Abadir, 1985) . Secondly since this method is overconstrainted. The synthesizer may not find a solution at all which meets the over constrained area and delay (ie. in OASIC, it may be integer infeasible). This approach avoids feedback after test incorporation through overconstraining.

It is interesting to note that there would be no test overheads at all for the architectural synthesized solution if test vectors could be generated such that the test constraints are met. Although automatic test pattern generators (ATPG) have been developed to handle combinational

(Roth, 1967) and sequential circuits (Marwedel, 1986, Agrawal, 1988) , these tasks require large computations in complex circuits. From experience (Goel, 1980) the computational complexity of test pattern generation has been shown to grow by the square of the gate count. For example it may take 77K cpu seconds to achieve 91% fault coverage for 1500 gates of a chip design (Agrawal, 1988) . If they were suitable to be used for complex VLSI chips and could meet the test constraints without changes to the hardware, then feedback to design synthesis would not be necessary and the test stage would only consist of additional time for ATPG to meet test constraints.

The final structured approach to the problem, choice 3, shown in figure 11.2, makes the best attempt to solve both the design and test problem. By integrating test incorporation into architectural synthesis a viable solution to the testable design problem is achieved through feedback providing testable design exploration. This approach is necessary for complex designs because functional testing is not practical nor sufficient and test vector generation alone is time consuming and very computationally demanding.

In the second case, the two main effects of test on architectural synthesis are:

1. The effect of test on the design cost (ie. as discussed in section 11.1.2 the testability of the design effects the cost of the chip, or the number of failed parts in the field)
2. The effect of test on design constraint satisfaction (ie. as discussed in section 11.1.4, the area and delay overhead).

In turn , the effect of the architectural synthesis on the test is

1. Complexity and size of the architecture effects the difficulty of the test incorporation.

2. The architecture also effects the difficulty of producing test vectors (ie. chaining operations may produce poorly testable functional units which may require a large number of test vectors).

Clearly design exploration and test exploration should be integrated in order to deal with complex chip design.

11.2 APPROACHES TO TESTABILITY

This section will outline the factors affecting the test (generation and evaluation)for a design . Test tools, such as fault simulation, automatic test generation, and controllability/observability tools, are generally discussed and referenced. Modifications to a design for testability will be discussed in section 11.2.2. Additional detailed information can be found in Proceedings of International Test Conference, Design Automation Conference, International Conference on Fault Tolerant Computing and the IEEE Transactions on Computers.

11.2.1 Test Measures and Tools

There are only approximate measures developed to indicate how testable a circuit design is. This first section will discuss the fault coverage measure, fault models, fault simulation, and test pattern generation very briefly to define terms and assumptions. The next section, 11.2.2, will cover design for test approaches to aid testing complex VLSI chips and discuss two popular methods in more detail.

Fault Coverage.

The most common test measure is fault coverage. Given a circuit represented by a specific design level, a set of fault models (suited to the design level), and a set of test vectors, the fault coverage is calculated by dividing the number of faults detected (multiplied by 100), obtained from fault simulation, by the total number of possible faults in the design.

This measure will vary according to the design representation level, the fault models, the test sets and the method of calculation. In some cases, discussed in the next section, it is not possible to calculate fault coverage and estimations are made. Table 11.2 illustrates the relationship between design levels, fault models, test tools(fault simulation and test pattern generation(TPG)), and some application areas where these techniques are common. The design representation levels for test measurement can be functional, gate, or switch levels , in order of increasing detail and accuracy. Different fault models can be used to represent faults at a node of the design. For example sa-0 refers to stuck at zero fault at a node.

Table 11.2. Test Options

Tools	Design Representation Levels		
	Switch	Gate	Functional
Fault Models	stuck at - 0,1,x,z, open,short	stuck at - 0,1,x,z	stuck at - 0,1 on I/O nodes or state transitions.
Fault Simulation	y	y	y
TPG	?	y:automatic	y:manually
C/O measures	?	C/Os	info-flow
Application	small circuits	most common	microprocessors

There are many tradeoffs involved in selecting a design representation level to use. The highest level is the functional representation which is technology independent. It is often used for microprocessor design because it can be implemented through test microcode which can be easily down loaded and executed on the chip. However it is the most inaccurate and difficult to measure test effectiveness. In fact studies have shown that a 90% functional level fault coverage may actually only corresponds to a 60% gate level fault coverage. Although the switch level provides the most accurate fault coverage measurement it is also the most computationally demanding. Most fault simulators use gate level representations for test vector generation. Gate level fault simulation is extremely computational even with concurrent execution (Agrawal, 1988) , and circuit size reduction techniques such as gate collapsing as discussed later in this section.

Fault Models.

The most common fault models include the single stuck-at (0,1,X,Z,open or closed) model. However not all failures can be modeled by stuck-at faults (Galiay, 1980) . The stuck at 0,1,X,Z faults are the most easiest to model and techniques of fault equivalence or fault collapsing can be used to decrease the total number of faults to consider. An example of fault collapsing is to represent a stuck at zero on an input to a nand gate as the equivalent of a stuck at one on the output of the nand gate since the fault has the same effect. Therefore only one of these two faults needs to be modeled. Stuck at open faults are more difficult since they essentially turn a combinational circuit into a sequential circuit. Bridging faults (more difficult to model due to layout information needed), delay-type faults (slow-to-rise,etc), or structure specific faults such as cross point faults in PLAs or neighboring faults in memory can also be modelled. Other faults include pattern sensitive faults and some strange faults which cause new transistor structures to form (Shen, 1985) for which fault models do not exist.

In order to reduce the number of fault models considered in a circuit some researchers (Shen, 1985, Galiay, 1980) have investigated the percentage and type of faults present in fabricated circuits. Shen found 43% (line or transistor) stuck at faults, 21% floating lines, and 30% bridging faults and 6 % miscellaneous (Shen, 1985) . Stuck at open were less than 3 % (Shen, 1985) . In another study (Galiay, 1980) the observed failure modes consisted of 55% shorts, 20% open faults and the remaining 25% were inobservable or insignificant.

Fault Simulation.

Many approaches to fault simulation have been investigated. In general, given a circuit with F faults and T test vectors, perform a fault simulation to determine the number of faults detected at the outputs of the circuit. F faults refers to the number of unique faults of the design, after fault equivalence techniques have been performed. The simulation requires F+1 circuit representations to be simulated concurrently. Each of the F circuits have one unique fault. The extra copy of the circuit represents the good circuit. As test vectors are applied at the circuit inputs, the circuit outputs are compared with the good circuit outputs. If the outputs differ then the fault is detectable. This algorithm continues until all faults have been detected or until all test vectors have been exhausted. The number of remaining circuits with undetected faults are summed and divided by F to determine the fault coverage. Serial, parallel, deductive and concurrent fault simulation algorithms have been implemented.

Various attempts have been made to decrease the complexity of fault coverage estimation. These include decreasing the types of fault models considered, decreasing the number of nodes to test for faults , and modifying the circuitry to ease fault detection or reduce the probability of faults being present. Another approach to decrease the number of nodes to test for faults other than gate /fault collapsing is to select faulty nodes using statistical methods.

Test Pattern Generation.

Automatic pattern generation for combinational or sequential circuits have been programmed. Most ATPG work in conjunction with a fault simulator to decrease the execution time. For example after a number of vectors are calculated, using a path sensitization algorithm , a fault simulation is run to delete other faults which are also detected by the new vectors (Marwedel, 1986) . Ten times speed improvement is achieved over conventional concurrent fault simulators, through fault reduction techniques. Automatic pattern generation for combinational units researched include (Roth, 1967) or other algorithms such as (Motohara, 1986) which is implemented on multiprocessor architecture. ATPGs for sequential circuits have been researched (Marwedel, 1986) or (Agrawal, 1988) . Both are used in conjunction with a concurrent fault simulator.

Hierarchical test pattern generators have also been developed with limited success. (Ho, 1984) illustrated this concept for testing a parametrized adder and (Varma, 1988) created a cell test generator and hierarchical test generator using Prolog for standard cells and iterative arrays. These approaches hold the most hope for integration of test pattern generation knowledge into parametrized silicon compiler cell libraries.

It has been shown that some controlability and observability (CO) measures (Goldstein, 1980) do not correlate well with testability of the design (Ratiu, 1982) . It has also been shown that a larger variety of COs do improve the performance of an ATPG, however no CO measure is superior over the others (Chandra, 1989) . CO at the functional level is called information flow. These measures have been used for functional measures of testability as in (Agrawal, 1980, Fung, 1982, Dussault, 1978) , and used in an automated test system in (Fung, 1986) .

Controllability/Observability.

Controllability observability tools have been developed as another means for evaluating the testability of designs without doing time consuming fault simulation and as an aid to ATPG by simplifying the backtracking process as discussed above. These measurements are automatically calculated for each node in the design. They identify nodes which will possibly be difficult to control (and therefore difficult to sensitize) or difficult to observe (and therefore difficult to propagate fault presence to outputs of the circuit). Both these tasks are important for testing the circuit.

In summary, given a specific design composed of combinational and sequential circuits, the testing phases (without hardware modifications) is computationally demanding, and very complex. The next section will outline how design modification can aid the test problem.

11.2.2 Design Modifications for Testability

The most common and popular method used to reduce the complexity of the test problem is to modify the circuit design for testability. There are generally two techniques to do this. One is the adhoc and the other is the structured approach (Williams, 1983, McCluskey, 1986) . These approaches do not avoid the problem of test generation and fault coverage, discussed in the previous section, however they do provide a method for dealing with the problem complexities through design partitioning. We will concentrate on discussing some structured design for test approaches. The structured approaches provide the best solution for incorporation test in VLSI chip designs.

The ad hoc approaches to test involve partitioning the circuit to gain access to smaller networks of the design. For example through multiplexing in and out internal nodes of an embedded circuit, additional control and observability is obtained. Bus based architectural designs also support the adhoc approach to test by providing partitions of the design and allowing access to several modules attached to the bus. This is very popular for testing microprocessors (Williams, 1983) in addition to the functional testing which uses test microcode described earlier. As long as the other modules outputs on the bus are in the high impedance state then the specific module can be isolated for testing using the bus for test vector transfer. The only problem with testing this design is the difficulty in detecting the cause of a fault on the bus.

The structured approach is recognized as the most suitable approach for complex chip, board, or system designs. In effect the circuit is modified so that all sequential circuitry, for example registers, are replaced by serial shiftable registers providing access to their storage elements by serially shifting vectors in and out of the chip. Only combinational circuitry remains which is now accessible through the serial shiftable test registers. This simplifies the circuitry to test since only a combinational test pattern generator is required for smaller partitions of the whole chip design. After ATPG is used on the extracted combinational logic islands, the test vectors are serialized and usable test vectors are created for the circuit.

Designs composed of a data path and FSM or micro code controller may require two separate testing methodologies due to their circuit differences as in (Fung, 1986) . More than one test methodology may be applied to the datapath as in (Abadir, 1985) , or different test methodologies could also be applied to the partitioned controllers as in (Fung, 1986) . This is called nonuniform test incorporation. An alternative is to apply the same test methodology to the whole data path as in (Krasniewski, 1985b, Craig, 1988) , called uniform test incorporation. The controller type test methodologies will not be discussed however more information can be found in (Abadir, 1985) .

Approaches in design for test differ in

1. how to partition the circuit,
2. with what test structures to replace the sequential registers,
3. how to implement circuit and test clocking, and
4. how to then calculate the test effectiveness or fault coverage. (Obviously the test overheads will also vary as briefly shown in table 11.3.)

We will outline two basic test structures, scan path and built-in self test, in the next section. We chose these due to their simplicity and popularity. Other structures include LSSD , counting ones , syndrome testing (Savir, 1980) , walsh spectrum and many other types of structures (Williams, 1983) used for scan or built-in test.

Scan Design.

The scan design test methodology (Williams, 1983) replaces sequential circuitry or registers with serial shiftable registers. Examples of some implementations of the test registers or the D type master slave flip flop are shown in figure 11.3 (Williams, 1983) .

By connecting the serial interconnect between the test registers, full controllability and observability of the memory elements is provided. Furthermore the test problem now becomes one of testing the remaining combinational circuits. During the test mode, each test vector must be shifted on chip, the system clocked once, and then the result of the combinational logic is then shifted off chip (simultaneous with shifting the next test pattern on chip). The test time is equal to the longest length of the scan chain multiplied by the number of test vectors. A full scan refers to changing all registers into scannable registers as in (Agrawal, 1984) . However not all registers need to be transformed into scannable registers in order to apply test vectors to all combinational

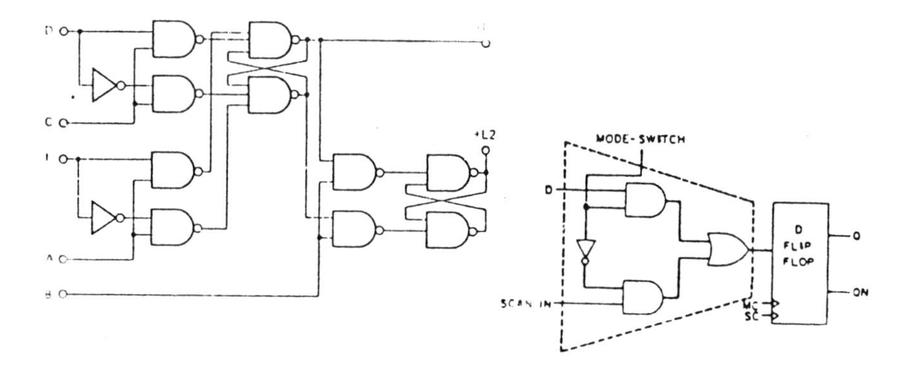

Figure 11.3. Examples a,b) of Scan register Design.

units. For example CO software could be used to determine which nodes should be controlled and observed in the scan registers as in (Fung, 1986) . For some applications additional scan registers could be used such as in the partitioning algorithm of (Funatsu, 1975) where combinational networks are separated into smaller networks to aid test pattern generation, using a back tracing algorithm.

More than one scan chain can be implemented to trade off test time with additional input and output pins. Thus test patterns can be shifted on and off the chip in parallel for more than one scan chain. Other similar techniques to the scan design exist, for example scan/set logic described in (Williams, 1983) where test scan registers are not part of the original design, and random access scan (Williams, 1983) where registers are accessed through addressing instead of through serial shifting.

Built In Self Test.

Built-in self test provides a methodology for automatically generating pseudo random test vectors (called TPG) on/off the chip and having the responses compacted (by a SA) on/off chip. There are three modes of behavior for the BIST registers which again replace the sequential circuitry. They are serial shifting, normal registers, and linear feedback registers (or TPG / SA mode). Essentially an initialization pattern (or seed) is shifted into the TPG and SA on the chip. Then the TPG starts to generate patterns and the SA compacts the combinational networks output responses. After T number of clock cycles the response is shifted off chip and the TPG/SAs are reinitialized for another testing period. Exhaustive testing, where T is the maximum sequence generated by the BIST structure, is often not practical (Wagner, 1987) , so pseudo-exhaustive testing may be performed. In the later case the pseudorandom numbers are generated for T clock cycles, where T is less than the maximal sequence. Analysis for fault coverage estimation and test confidence prediction have been researched (Wagner, 1987) for this case.

The most common TPG or SA circuit is the LFSR which can be transformed from master slave D flipflops. An example is given in figures 11.4 of a LFSR register. Other structures have also been examined for use as a pattern generator or signature analyzer, such as counting techniques for SA and cellular automata for TPG. The differences of these structures lie in their area overhead, delay overhead, and fault coverage estimation.

Again the chip is transformed so that smaller combinational subnetworks are being tested. BIST with TPG on chip has the advantage over scan path design for test of testing the chip at speed, instead of clocking once in the scan method to obtain one response at a time. However the LFSR has a larger area overhead than the scan register, especially if each combinational network has its own separate TPG and SA. For designs

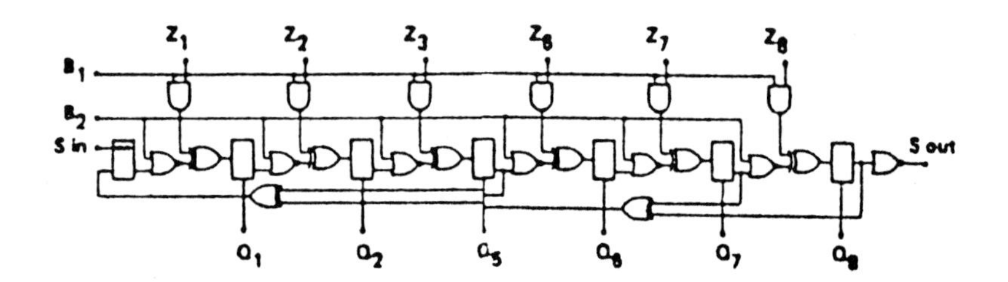

Figure 11.4. An Example of a LFSR Design for BIST.

that do not have scan paths, TPG and SAs may be added through multiplexors. The large area overhead may be avoided by using multiplexors to share the TPG among more than one combinational network. However the response compaction or signature analyzer circuits must be separate for each functional unit concurrently being tested. Another implementation of BIST allows scheduling combinational units for testing at different times, called test phases. This method is often called BILBO, built in logic block observer (McCluskey, 1986, Williams, 1983) . This is normally implemented by translating a scan path is directly translated into BIST circuitry and 1/2 the CUs are tested in one of two phases for orderly circuits (McCluskey, 1986) . In this implementation more than one CU may be tested at the same time. Thus LFSRs will be created from existing registers and fed to CU inputs and from CU outputs separately. However a LFSR used as TPG in one phase may be used as a SA in another phase.

Measures of Test Overhead.

The earlier DFT is implemented the better the designer can judge design overheads due to test. This provides earlier estimation of whether the final design will meet design constraints or whether the design will be testable enough to satisfy test constraints. In either case when this is known the designer can find another solution or continue finishing the current design for fabrication. Table 11.3 illustrates some figures for overheads determined from experience of various researchers. Overheads vary greatly according to the design, technology, and the test implementation.

Table 11.3. Some Quoted Overheads for ScanPath

References	Overheads	
	Area %increase	Speed %decrease
(Agrawal 1984)	~12	~10
(Williams 1983)	4-20	small
(Susskind 1984)	5-15	"

The test approaches traditionally were viewed as a continuous effort to increase testability of the design through modifications and fault coverage estimation. No feedback to the original designers intent or to design synthesis was examined apart from general layout rules (Galiay, 1980) to decrease chances of process errors causing faults, and the adhoc and structured methods to modify the circuit to avoid sequential circuit testing (Williams, 1983) or to increase ease of generating test vectors (Bhatt, 1986) .

11.3 PREVIOUS RESEARCH IN DESIGN FOR TEST

We will concentrate on a discussion of previous research in structured design for test automation for digital designs. A large amount of research in ATPG, new test structures, new fault models, controllability and observability and other areas in the test field were referenced and outlined in the previous section. In this section reference to these tools only in the design for test (DFT) environment will be discussed. DFT research falls into two categories. DFT for special structures and for general purpose digital systems. Each will be discussed below.

Research in automating test decisions for special structures, such as PLAs, have demonstrated that the selection of how to test a circuit is not trivial, and nor are the tradeoffs clearly defined (Zhu, 1988) . No approach is clearly better and searching is often required to determine the best test approach given an application (Zhu, 1988) .

Research in automating ad hoc design for test techniques have been investigated. For example in (Chen, 1985) controllability and observability measures are used to determine which nodes of a circuit should become primary inputs or primary outputs. Small circuits were used to demonstrate this techniques, however this has been extended to determine which nodes should become part of the scan chain in (Fung, 1986) instead of transforming them into primary inputs or outputs. Another example of automating ad hoc techniques is in (Bhatt, 1986) where combinational logic circuits are partitioned for improved testability. Block timing is maintained and modifications are made so that every output node depends on a small number of input nodes.

Higher level automated methods of implementing structured design for test have also been investigated. Some examples of these are automated scan path design in TITUS (Agrawal, 1984) , automated test for finite state machines in Silc (Fung, 1986) , automated nonuniform test for architectural design in TDES (Abadir, 1985) or for finite state machine design (Zhu, 1988) , or extensions for BIST automation in (Craig, 1988) and BEST (Built in Exhaustive selft test) automation in (Krasniewski, 1985b) . The Silc and TDES systems will be discussed in further detail in section 11.4 because they also involve design synthesis in some form.

TITUS (Agrawal, 1984) provides automated scan path incorporation for custom polycell based designs. All flipflops are connected into shiftable registers during testmode so that the automatic test pattern generator for combinational logic can be used. Better delay optimization was observed by interconnecting the scan registers after layout.

Silc (Fung, 1986) involves an integrated system that includes testability rules, testability expert, test structures and a testability evaluator. The input to the system is a description of a number of finite state machines. The testability evaluator uses (a) information theory (Agrawal, 1980) to identify hard to test finite state machines, (b) controllability observability measures at the structural level, and (c) a path tracing technique to identify critical testing paths (ie components for scan testing). The testability expert makes the decisions about which test methodology to apply to a finite state machine in the design based upon the testability evaluator output, user requirements, and the test structure attributes.

The TDES (Abadir, 1985) incorporates non-uniform test methodologies into a circuit design based upon the combination that gives the best multiple-criteria score of test attributes. The test attributes includes area overhead, test execution time, possibility of sharing BIT structures, amount of circuit tested for free, fault coverage, I/O requirements, external test equipment requirements, and need for test generation. A structural design solution is used where kernels or combinational logic (requiring test) and interconnect paths, or I-paths (which may include busses, muxes or registers also), are identified in a graph data structure.

When the kernels cannot be embedded with known test templates, additional circuitry is added to modify the kernel to allow testing. Additionally when an interconnect path is needed for testing by two different kernels, this forces the two kernels to be in different test phases or can be solved using test steps as in (Abadir, 1985) .

The other systems (Craig, 1988, Krasniewski, 1985b) provides uniform test incorporation of BILBO modules into the circuitry. In (Krasniewski, 1985b) the problem of test scheduling to minimize the test phases without adding new interconnect paths is investigated. In (Craig, 1988) test control architectures were also investigated, using a star, bus, and multiple bus configuration for control.

11.4 APPROACHES TO TEST WITH SYNTHESIS

Although many researchers stress that testability should be considered during the early stages of design (Williams, 1983, McCluskey, 1986) , most testability research has been done after a structural design solution is defined with no feedback to the original synthesis process for finding more testable designs. Some testability research work relevant to our problem is outlined below.

11.4.1 Previous Research

Testability incorporation in the Silc silicon compiler (Fung, 1986) utilizes testability measures at the functional, and structural levels to guide test incorporation. The functional measures use information flow analysis, to group the finite state machines and incorporate testability. The structural testability involves calculating controlability and observability measures for circuit nodes of the data path. However no feedback to the synthesis process is provided and no other test methodology for the data path is considered apart from the scan path inclusion.

The TDES, or testability design expert system (Abadir, 1985) implements testability in a graph-based structural design by matching the subcircuits requiring testability with test design templates. The test design templates contain measures of test time, area overhead and the control sequence for a particular test strategy. Again no feedback to a synthesis process is used and the test implementation is based on local structural enhancements with no global information.

Built in exhaustive self test incorporation in data paths has been investigated at the University of Rochester (Krasniewski, 1985b) . Built in self test (BIST) module selection, placement, scan path organization, mode controller organization, and derivation of test procedure is handled by the software. Although speed estimates have been produced, no design synthesis or feedback is provided.

11.4.2 Commercial Systems

Silicon Compiler Systems Inc (Sabo, 1986, Johannsen, 1987) Genesil structural cell compiler provides three controls for the designer wishing to incorporate test into their design. They are none, full, or partial test visibility. These refer to the number of registers in the design which will be transformed into scannable registers for a scan path or LSSD test implementation. The option full refers to using all shiftable test latches in the design. The partial visibility requires the user to specify the sequential depth limit which is used to select the registers to be transformed into scan registers for testing. Built in self testing is also provided by linear feedback shift registers, (Sabo, 1986) which are provided as an additional configuration of the shiftable test latch in the compilers library.

Silc Technologies Inc's (Rosales, 1989) newer tool provides automatic test incorporation for synthesized designs by transforming registers in feedback paths into scannable registers and ensuring other register are inactive during test mode (clock and reset signals are inactive). This transforms a design into a combinational logic design model which allows them to run an ATPG to produce test patterns for the design. Scan facilities are added automatically, and clocking control logic for the test circuitry is also added. The industrial tool described above is quite different from the original research discussed in the previous section (Fung, 1986) .

11.5 INADEQUACIES OF CURRENT SYNTHESIS WITH TEST

There are remaining problems with tools in both the synthesis domain and the test domain which were discussed in section 11.4 and 11.3. We will discuss the problems with systems developed or researched as a means of integrating both design and test tools. These problems can be classified into three main categories each discussed below. The categories are feedback, integration, and constraint estimation.

11.5.1 Feedback

Proposed systems which integrate design synthesis with test incorporation lack the capability of providing feedback to architectural synthesis after test incorporation. This is an important means for exploring the testable design domain. For example if the test overheads cause the design constraints to be exceeded, then a solution cannot be found without feedback. If the user is to provide feedback, then it is very difficult to determine what changes are to be made by the user to the synthesizer to produce better design solutions which will have lower test overheads. Since the testable design exploration is a very difficult task, automated feedback for synthesis exploration methods must be created.

11.5.2 Integration

A lack of integration between the synthesis and test tools also causes problems for testable design synthesis. Apart from difficulties of implementing automated feedback, the absence of integration causes isolated decisions to be made in synthesis in the absence of its effect on the test process. These isolated decisions may result in poor design solutions. The lack of integration in these two cases is largely due to the fact that the tools are produced by two different group of researchers. An unnecessary duplication of the data base is often created further hindering design automation tools. By considering integrated design and test synthesis simultaneously the overall constraints may be better satisfied and better solutions may be synthesized.

11.5.3 Constraint Estimations

Unfortunately there is a lack of standards in the area of constraint estimation. However the separation of synthesis constraints from testability constraints causes more problems during testable design exploration. Since few synthesizers even consider area estimations or floorplanning, it makes the problem even more difficult to estimate the overhead in area (and speed) due to test. Design and test tools must create area, delay and test constraint estimations using the same methods. Standards should be created for this purpose.

Clearly these three problems (feedback, integration, and constraint estimation) are very important for further enhancing and improving architectural synthesis tools. One approach to solving the problems outlined above is presented in the next chapter.

12.

THE CATREE ARCHITECTURAL SYNTHESIS WITH TESTABILITY

Two VLSI testable architectural synthesis methodologies with testability, area, and delay constraints are presented in this chapter. This research differs from other synthesizers by

1) implementing testability as part of the synthesized VLSI architectural solution,
2) providing feedback to the synthesis process, and
3) by integrating test incorporation with architectural synthesis (specifically allocation and binding) using a binary tree data structure.

These design and test synthesis approaches are vital to the acceptance of synthesis tools in industry by providing feedback to the synthesis search when constraints cannot be met. Furthermore they will help to decrease the VLSI design cycle times by considering test constraints early in the design.

Both testable design synthesis methodologies are presented in this chapter. The testable design synthesis algorithms are discussed and synthesized examples help illustrate the techniques. Results from the first methodology, CATREE, show that the 'best' testable design solution is not always the same as that obtained from the 'best' design solution of an area and delay based synthesis search. Preliminary results of the second methodology, CATREE2, indicate that better design solutions are obtained by incorporating test during design synthesis as opposed to approaches which incorporate test after a structural design solution is formed.

12.1 PROBLEM DESCRIPTION

We propose a solution to the following problem.

> Given a general algorithmic description of a behavior with area, delay, and test constraints, perform a datapath design synthesis by mapping the algorithm into a chip design which satisfies the given constraints.

If we assume that OASIC is used to provide an initial optimized schedule and allocation of hardware (before test incorporation) the problem then becomes the following. Given a scheduled DAG, perform allocation and binding so that a testable architecture is synthesized.

Our approach to solving both of these problems is called CATREE (for Computer Aided TREEs), and will be presented in the remainder of this text. For CATREE the choice of test methodology is explored to further search for a design solution. In CATREE2, one specific test methodology is incorporated. Both approaches will create a testable design using area, speed, and testability estimates to guide the search through the design space. Our testable design synthesis methodology only considers synthesis of the data path, and not the controller. However a control table is output from the synthesizer which could eventually be translated to interface to finite state machine controller synthesizers as in (Wei, 1987) and also incorporated with test as in (Abadir, 1985) . Design solutions use a two phased clock with master-slave registers as described in part II and III (OASIC). All functional units are assumed to be combinational logic.

Design constraints include area, delay and testability estimates of the synthesized data path, further discussed in section 12.4.3 and 12.4.5. The area constraint includes the areas due to the hardware components and the interconnect of the architecture. CATREE uses a binary tree data structure and heuristic algorithm which minimize resources. The two dimensional characteristic of the binary tree data structure aids the area estimations. The circuit delay refers to the period of the system clock (or inverse of the clock speed) and is more refined than the delay in OASIC (which is an integer representing the number of clock periods). Our delay model is similar to (McFarland, 1986) where we include delays through registers, multiplexors, and functional units. Also the delay due to the interconnect length, and fanout is used.

We define the testability constraint as a measure of the estimated fault coverage (estimated number of faults detected divided by the total number of faults in the chip) and the test time (or the number of clock cycles required to test the chip). We assume the synthesized chip design will have a test mode which is externally controlled by one pin. Our test models, further discussed in section 12.4.4, are a x-chained scan path, x-phased BIST, a shared BIST implementation , or a combination of these methods. For a scan path implementation (Williams, 1983, McCluskey, 1986) , test vectors are supplied externally and serially shifted on chip using one or more scan chains. For a BIST implementation (Williams, 1983, McCluskey, 1986) , the test generator

(LFSR) is located on the chip along with the signature analyzer. LFSR initialization seeds and signatures are loaded on and off the chip by serial shifting them through a single scan chain.

The output or the synthesized architecture from CATREE is composed of a floorplan, netlist of hardware resources, and a mapping of code operations and variables to these hardware resources. These resources are registers, interconnect, buses, multiplexors, and functional units. The mappings of code operations to functional units and the mappings of the variables to registers and busses are output. The execution times of the code operations in the functional units and lifetimes of the variables in their registers are also given, as are the variable transfer-times through interconnect and hardware components are output.

Our testable design synthesis methodologies will search through various designs until a solution that satisfies the design constraints is found. The tool does not continue searching for a better solution once a design solution is found. However if a design solution cannot be found (for example due to an overconstrained specification) then the 'best' solution found by the tool is output. Solutions are judged by a multiple-criteria performance measure to be discussed in section 12.5.

The CATREE(2) methodologies can be applied to ASIC, cell/silicon compilers or full custom designs approaches. The customization for different cell libraries is done through the synthesizers library file. The library file contains the types of functional units supported (defined as a list of operations which it can perform), their propagation delay (from input data to output data) the delay between successive input data, and the area (or width and length) of the functional unit. After the synthesized testable design solution is produced, its netlist and floorplan is output. This could then be interfaced to a specific program for the final placement, routing and layout, such as (Bhandari, 1988) , as shown in figure 1.2 of chapter 1, to complete the VLSI design. The design would then be ready for fabrication. In gate arrays or standard cell methodologies macro cells could be built out of various gates in the libraries and then placed and routed with the other components as in (Pangrle, 1987) .

The library file also contains testability information for each functional unit. We assume the functional units have been precharacterized for test, In other words one can assume that an automatic test pattern generator has previously been run on the library cells to characterize them for testability. In the future this could be extended to parameterizing the test characteristics (similar to parameterizing a module with 16 bit or 32 bit data width inputs for generating the layout) of modules as suggested in (Fung, 1986) . This test precharacterization provides fast testability estimation, described in detail in section 12.4, and saves time by avoiding regeneration of test patterns to estimate the testability for each new design during the synthesis exploration. The test data, stored for each functional unit in the library file, is shown in table 12.1 below. Different test sets may also be stored for each functional unit. For example a test set with 5000 vectors that achieves 90% fault coverage and another test set with 1700 vectors that achieves 80% fault coverage could be stored.

Table 12.1 Test data for each functional unit in library file.

	Total # of faults of hardware unit
Scan Path	The test vectors (from ATPG) Total # of test vectors Total # of faults detected
BIST	The polynomial and initial seed The length of pseudo-random # sequence Total # of faults detected

12.2 COMPARISON WITH PREVIOUS RESEARCH

Most research on architectural synthesis has not included testability incorporation. Only estimates of area and delay have been examined to provide feedback into the design search (McFarland, 1986) . The DAA (Thomas, 1983) , FACET (Tseng, 1986) , BUD (McFarland, 1986) and other approaches (McFarland, 1988) provide architectural synthesis, however, no test incorporation is performed.

Although many researchers stress that testability should be considered during the early stages of design (Fung, 1986) , most testability research has been done after a structural design solution is defined (Craig, 1988, Krasniewski, 1985b, Abadir, 1985) with no feedback (Fung, 1986) to the original synthesis process for finding more testable designs.

For simplicity CATREE uses a scan path or a BIST design for test methodology (Williams, 1983, McCluskey, 1986) to implement testability. CATREE implements the test methodology so that all functional units are tested thus avoiding calculation of controllability and observability measures (Goldstein, 1980, Ratiu, 1982) for selection of nodes to test. However use of these controllability and observability measures may decrease test overheads as discussed in section 12.6.

Silc (Fung, 1986) provides automatic test incorporation; however, no feedback to the synthesis process is provided and no other test methodology for the data path is considered apart from the scan path inclusion. Silc was discussed in further detail in chapter 11.

The testability design expert system, TDES (Abadir, 1985, Abadir, 1985, Zhu, 1988) , implements testability in a graph-based structural design; however, no feedback to a synthesis process is used and the test implementation is based on local structural enhancements with no global information. It attempts to implement a number of test methodologies for combinational logic blocks or kernels that fit into existing interconnection structures. TDES may try different test methodologies on a design but may schedule them separately or concurrently depending upon the best test execution time and overall design score amongst other test embedded alternatives. TDES is driven bottom up by selecting the best test methodology for each kernel or part of the design.

12.3 TWO SYNTHESIS WITH TEST METHODOLOGIES: CATREE & CATREE2

Our first testable design synthesis methodology, CATREE (Gebotys, 1987, Gebotys, 1988a, Gebotys, 1988c, Gebotys, 1989) (for Computer Aided TREEs), enhances the state-of-the-art in the area of VLSI design synthesis with testability constraints by including the following features.

- Testability is implemented as part of the VLSI architectural solution. Testability, area, and delay estimates are used to guide the design synthesis search.
- A two dimensional binary tree data structure (McQueen, 1984) is used throughout architectural allocation, binding and testability incorporation. Design hierarchy, partitioning, and two-dimensionality naturally represented with the data structure are used to advantage for design solution searches, constraint estimation, and test methodology incorporation.
- This design and test methodology provides a larger, more complete, and flexible design search.

CATREE allows the exploration of the effects of different test methodologies on a specific design, and the effects of a specific test methodology on different design solutions. CATREE uses nonuniform test incorporation as a means of concurrently testing two or more different design partitions. It is also driven top down by the selection of test methodologies to incorporate. The CATREE design synthesis with testability constraints approach is shown in figure 12.1. The circles represent the constraint estimation (area, delay, test), whereas the squares represent the tasks being performed.

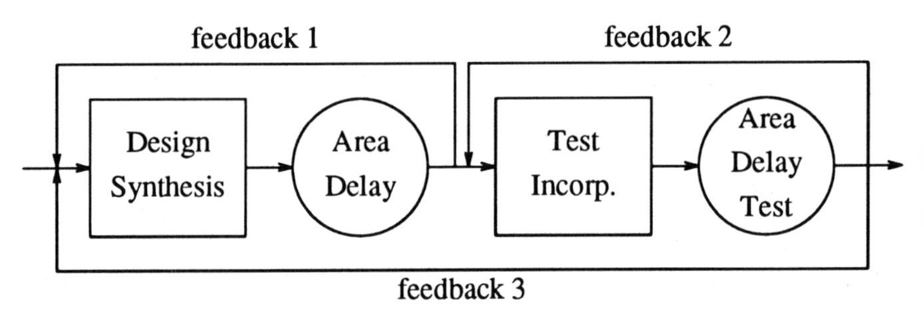

Figure 12.1 The CATREE VLSI design methodology with testability constraints. Test is incorporated after the design is synthesized.

Our second testable design synthesis methodology, CATREE2 (Gebotys, 1988b) (for Computer Aided TREEs version 2), an extension of CATREE, provides simultaneous design with test synthesis, satisfying test, area, and delay constraints. The test methodology to be incorporated is fully specified by the user. Highlights of the CATREE2 design and test methodology, not found in CATREE (Gebotys, 1988a, Gebotys, 1988c, Gebotys, 1987) and previous research (McFarland, 1988) , include the following.

- Simultaneous architectural design and test synthesis allocation and binding algorithms. Early design decisions are based upon normal and test mode behavior.

- Single feedback design synthesis methodology based on area, delay, and testability constraint estimations to guide the design search.
- Different sets of weights for cluster rules are used for exploring design tradeoffs.

The CATREE2 integrated design and test methodology is shown in figure 12.2. Finish design allocation refers to performing register, test register, and bus allocation.

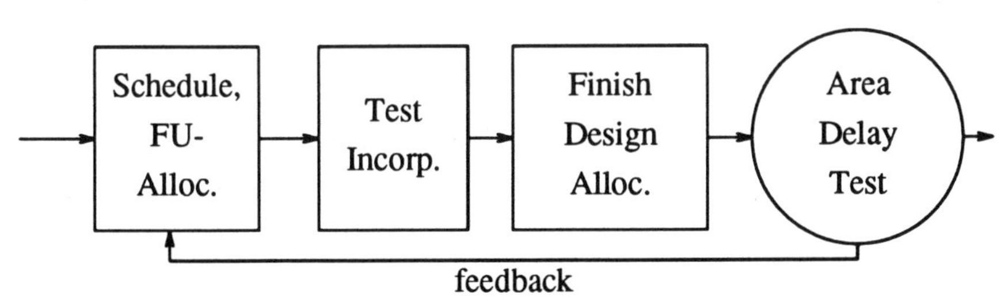

Figure 12.2. CATREE2 integrated design and test synthesis methodology. Design and test information used to make design synthesis decisions. (FU-Alloc refers to functional unit allocation).

CATREE and CATREE2 use the same binary tree data structure to integrate design and test synthesis. CATREE2 uses different and improved synthesis algorithms which consider both design and test information. It was developed after CATREE. CATREE2 is applicative to top-down system design where the test methodology is constrained from higher level design decisions.

Both methodologies are vital to the acceptance of synthesis tools in industry by providing automatic feedback to the synthesis process when the design constraints can not be met due to test overheads. Otherwise the design problem is left unsolved with the emphasis on the user to determine what has been done by the synthesizer and how one may reinvoke it to provide a different solution which will hopefully have lower test overheads.

Both CATREE and CATREE2 testable design synthesizers will be discussed in this chapter. Examples will be given to illustrate their algorithms. Results from the testable design synthesizers are presented and discussed. CATREE is presented in section 12.4 and CATREE2 in section 12.7. A comparison of both approaches is given in section 12.8. Since the CATREE and CATREE2 methodologies schedule first and then perform allocation and binding of (test) registers, busses and bus drivers, one can use the OASIC synthesizer (presented in Part III) to provide an optimal schedule for input into CATREE. This is also important because it immediately provides a minimum number of resources, from which test overheads can be calculated at the end of the CATREE methodology.

12.4 CATREE DESIGN SYNTHESIS STAGES

The CATREE synthesis stage consists of parsing the input specification, and binding of functional units, registers, and interconnect. OASIC is used to obtain an optimized schedule and functional unit allocation, as well as an initial allocation of registers and busses from which over heads can be calculated. No test methodology consideration is made during these design synthesis algorithms.

Since the design for test problems are NP-hard (McFarland, 1988, Craig, 1988) , and therefore it is likely that no optimal solution can be found, we've attempted to solve these using heuristic algorithms. The allocation algorithms attempt to produce solutions with a minimum number of long interconnects which is important for future integrated circuit designs (Holton, 1986) . Each algorithm will be briefly discussed below.

12.4.1 Input Specification

The input specification, written by the user, provides the algorithmic code sequence, design constraint specifications, design library identification, initial schedule name, and a list of test methodologies to explore. Figure 12.3 gives an example of the input specification. Only straight line code segments are synthesized in CATREE for simplicity. Unlike OASIC a language was chosen instead of a data-flow graph because the interface (or parser) was faster to code and design constraint specification was also easily supported. Also mathematical and scientific applications for custom VLSI design synthesis are most easily found in algorithmic form (Trickey, 1987) . The user may specify extra constraints in the code sequence, using labels and arcs. The labels and arcs are used to force an operation to be executed before another operation. In particular this allows the user to examine tradeoffs between the cycle time and the number of functional units which will be discussed further in section 12.4.3. OASIC performs the scheduling as described in part III.

12.4.2 Design Allocation

After scheduling and initial allocation in OASIC, the schedule is transferred to CATREE. The binary tree data structure (McQueen, 1984) is used throughout CATREE including: design binding, test incorporation (test binding and test allocation), and constraint estimation stages. The binary tree data structure provides the following three characteristics important for design synthesis. First, the partitioning characteristic provides a solution to handling design complexity by dividing the large problem into smaller problems to solve. The second characteristic is the two-dimensional which is used for constraint estimation, test incorporation, floorplanning and biasing solutions towards minimum number of long interconnect. Finally the tree data structure has very simple and easily codable algorithms.

```
module          algo( inout : a,b,y,x,d: byte;
                in a,b,y,x : byte;
                out d : byte );
var             a,b,f,x,y,j,k,l,d : byte;
constraints     area = 2000, delay = 500,
                fault_coverage = 95, test_length = 20000;
library         generic;
schedule        asap;
test_method     bist, scanpath;
label           1,2;
arc             1 before 2;
begin
                f:= a + b;
1:              k:= f - x;
2:              j:= f * y;
                l:= f + j;
                d:= k / l
end.
```

Figure 12.3. A simple example of an algorithmic input specification for CATREE.

Each node of the binary tree data structure has two son-nodes and one father-node. Root and leaf nodes are the exception. For example the root node has no father-node and the leaf nodes have no son-nodes. The root of the tree will be referred to as the top of the tree for our terminology. We will refer to top down and bottom up tree algorithms defined as moving from the root node down the tree and from the leaf-node up to the root-node respectively.

The tree is formed by placing operations into leaf-nodes of a tree using a heuristic tree formation algorithm. The operation and its input and output variables are stored at each leaf node. Scores are computed for all pairs of operations. The score is the number of common variables used as inputs or outputs of both operations. A list of operations in order of high to low total sum of scores is formed. The two operations, first on the list (with the highest scores), form the initial tree. Then each of the other operations (in order of high to low sum of weights) is placed in the tree closest to existing operations in the tree to which they're most connected. Generally operations which have a large number of common variables are placed close together in the tree. Since nodes will be swapped or moved and merged during the design synthesis search, the simple tree formation algorithm appears to be sufficient. The tree algorithms attempt to decrease the complexity of the allocation algorithms by decreasing the wide range of cluster group choices. This is done by the partitioning of operations in the binary tree. The synthesized algorithms are biased towards solutions with minimum interconnect without sacrificing quality. The functional unit tree is also maintained through the methodology to ease functional unit searching during feedback as discussed in section 12.4.5.

Functional Unit Binding.

A bottom up tree traversal algorithm (Gebotys, 1988b) collects operations from the leaf nodes with nonconflicting firetimes and valid functional unit representation. For example an operation may require two cycles to calculate its output. These two firetimes must be different from the firetimes of all other operations that it will be merged with. Also the group of operations, for example (+,-,>=,<), which produces a functional unit, for example an ALU, are stored in a library file which is checked during functional unit binding. The functional unit in the library that has the minimum amount of functionality required is chosen. This algorithm works bottom-up so that operations which are close to each other in the tree will be merged into one functional unit. functional units close to each other in the tree shared variables and therefore will help to reduce the future interconnections in the design as discussed during register and interconnect binding sections.

The bottom up tree algorithm is approximately O(N), for small N, where N is the number of leaf nodes of the tree or operations in the algorithm. In general these tree algorithms are fast and easy to code. An example of this algorithm is shown in figure 12.4. The operation and its variables are stored at each leaf node and the functional unit is identified at the tree node containing the subtree of its operations. The functional unit node in the binary tree represents the future use of design area.

Reclustering of operation leaves can be performed to further minimize the number of functional units until the number of functional units is equivalent to OASICs. An example is shown in figure 12.5. The flexible data structure allows exploring different functional unit configurations for a particular schedule. Thus given a schedule we can explore functional unit allocations by reclustering the binary tree.

Register Binding.

Register binding uses a bottom-up tree traversal algorithm with variable cluster rules activated at each node of the tree. The bottom up algorithm is used so that variables will be clustered together first from adjacent highly connected functional units in the tree. This in effect will bias register allocation towards local registers (formed near functional units) and uniform registers (allocated throughout the tree or design area) amongst functional units.

The output variable of each operation is placed in a list and propagated up from the leaf nodes. Generally, variables or clusters of variables are merged into one cluster if their lifetimes do not overlap. An

```
%functional unit binding
%    highest nodes with the list of operations such that
%    no firetime conflicts among operations and
%    function exists in library
etc...
fu_alloc(   [L,Root,R], [L1,[Root|[FU]],R1], FU)<-
    fu_alloc( L, L1, Op_list_L),
    fu_alloc( R, R1, Op_list_R),
    can_form_fu( Op_list_L, Op_list_R, FU);
%
etc...
can_form_fu(   Ops1, Ops2, FU_lib)<-
    no_firetime_conflicts( Ops1, Ops2, Fu),
    fu_exists_in_lib( Fu, FU_lib);
```

Figure 12.4. Bottom up tree algorithm to perform functional unit binding.

example of variable clustering is shown in figure 12.6. Only output variables are clustered since each will be found at only one leaf in the tree (since they are defined only once in the algorithm). Input variables of operations at the leaf nodes were not clustered locally in this algorithm because each could be located at more than one leaf node of the tree and therefore allocated to more than one register. At the root of the tree the clustered list is merged one last time with a list of input only variables.

Each register, represented by a cluster of variables, is placed in a new leaf node closest to the functional units to which it is most connected. Register leaf nodes represent the future use of design area. Constants are placed in tree nodes in the same manner.

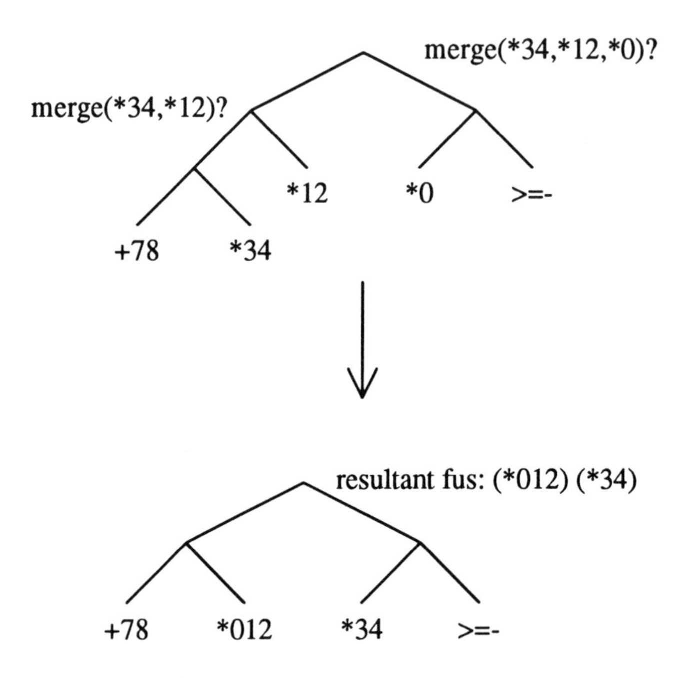

Figure 12.5. Reclustering of functional units (fus), using a bottom up tree traversal, to facilitate functional unit minimization or feedback. In this example, (three) multipliers are collected for minimization (into two).

Examples have shown that this algorithm can produce the minimum number of registers (Gebotys, 1988b) . This provides reasonable results for the test incorporation described in section 12.4.4.

Interconnect Binding and Allocation.

The objective of bus allocation is to minimize the number and lengths of the busses. This is more refined than OASIC, since the floor-planning information is now used and the length of the busses is important. The objective of interconnect allocation for random topologies, is to minimize the total number of inputs to the multiplexors, which may be

```
%Register Binding Algorithm Using
%Bottom Up Tree Traversal
%
register_alloc( [L,Root,R], Register_cluster_list) <-
   register_alloc( L, Variable_cluster_list1),
   register_alloc( R, Variable_cluster_list2),
   cluster_variables( Variable_cluster_list1,
      Variable_cluster_list2, Register_cluster_list);
%...etc
```

Figure 12.6. Part of register binding algorithm,(written in Waterloo Prolog) showing bottom up tree traversal.

located at inputs to functional units or registers. The allocation involves definition of cross variables in the tree structure, allocating interconnect, and minimizing the number of multiplexor inputs.

Variables transferred between registers and functional units, called cross-variables, are recorded at nodes in the tree. The node, where each cross variable is stored, is defined as having each son-node, or subtree, hold either the register or the functional unit involved in the transfer.

A top down tree algorithm allocates interconnect for a bus or random implementation by using cross-variable cluster rules activated at each tree node. A top down algorithm is used so that cross-variables at high nodes of the tree are first clustered together to minimize the interconnect at the root. There are four cluster rules which are listed below in table 12.2.

In each rule a check is made to ensure that there are no time conflicts between the cross variable transfers (unless it is the same cross variable). Also a check is made to ensure that they are allowed to be allocated to

Table 12.2. Interconnect Rules

Rule #	Rule description for merging two cross variables
1	equal destination and equal source
2	equal source
3	equal destination
4	no time conflict

the same side of the functional unit. The interconnect allocated is then stored at the lowest possible node in the tree such that the subtrees below the node will contain all components using that interconnect. Figure 12.7 illustrate the top down interconnect allocation algorithm and its relationship to the interconnect located in the floorplan.

A heuristic local multiplexor minimization algorithm, which swaps cross variables of commutative operations between clusters to minimize the number of multiplexor inputs, was also used. The algorithm attempts to reduce the inputs of multiplexors by swapping inputs of a functional unit whose operation is commutative. Each functional unit is examined to see if common variables exist on both sides of the functional unit. If so, they are swapped in an attempt to minimize the overall number of multiplexor inputs. Only single swaps are tried. The top down interconnect algorithm attempts to minimize the number of long interconnect which are most likely to be located at high levels (such as the root node) in the tree.

12.4.3 CATREE Area and Delay Estimates

Area and delay estimations (Gebotys, 1987) are obtained by creating a floorplan, performing a bottom up area estimate, and then extracting delays of circuit paths defined in the tree. The objective is to determine if the synthesized design solution meets the constraints before test incorporation. This step was added to save time in the design exploration by

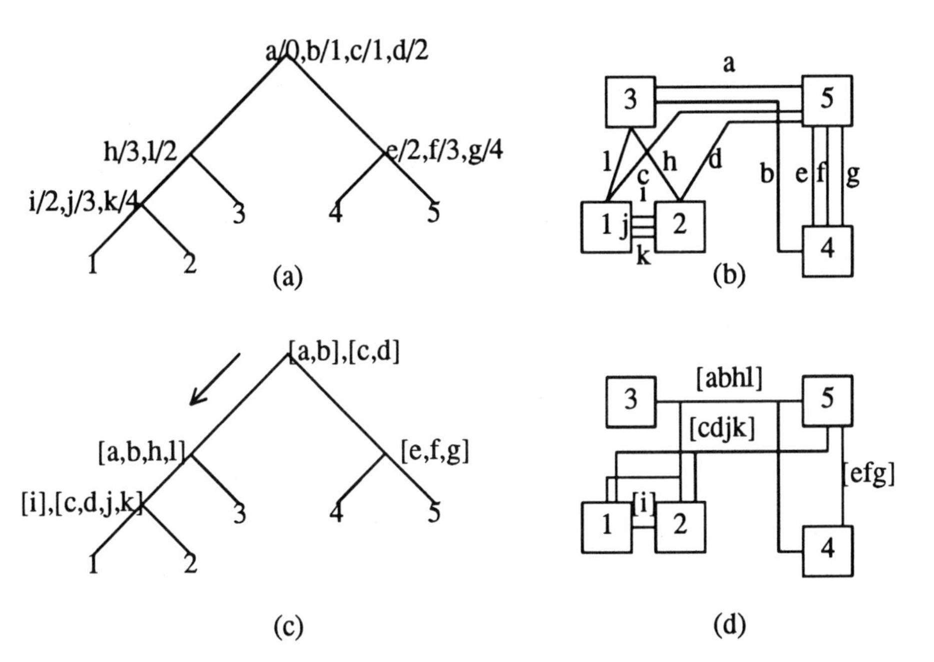

Figure 12.7. Interconnect allocation using top down tree algorithm to minimize the number of long interconnect. In a), cross variables and their transfer times (var/time) located at various nodes in the tree are shown. (b) shows the corresponding variable transfer paths in the floorplan (c) shows the top down algorithm clustering cross variables into 4 interconnect. (d) illustrates the solution in the floorplan. Only two long (top level, [abgl] and [cdjk]) interconnections are allocated. (Note: 1,2,3,4,5 can be a register or fu).

omitting test incorporation on designs that already do not meet the design constraints before test. A simplified floorplanning algorithm and constraint estimation technique is described below, to guide our prototype tool. However more sophisticated techniques as described in section 12.6 could enhance our results.

The simple heuristic floorplan algorithm moves top down through the tree alternatively assigning X and Y split dimensions. It also assigns low or high subtrees to each node attempting to place nodes close to their neighbors to which they are most connected across each dimension. The problem of placement or floorplanning for custom designs is very difficult as discussed in chapter 5.

An estimate of area and delay, similar to BUD (McFarland, 1986), is calculated by using a bottom up tree traversal algorithm. At each tree node the propagated minimum bounding boxes or areas of functional units, interconnect, multiplexors and registers are combined or summed. Delay estimates are calculated by outputting paths through the design, consisting of functional units, fanout, registers, multiplexors and interconnect lengths, and computing their delays. The interconnect paths, or wires attaching a source to a destination, are defined as following the split lines encountered in the tree, in the direction of the destination. An example in Figure 12.7 and the equations below further describe the area and delay estimations.

If estimates of area and delay do not meet the constraints, resynthesis is invoked by rescheduling operations and reallocating functional units. When the area constraint is not met, a bottom up tree traversal algorithm is used to collect functional units which can be merged together by rescheduling their conflicting operations. Also the user may selectively choose which functional units are to be merged or split if for example delay is not met. The 'best' solutions are stored in case no further beneficial merging can occur.

12.4.4 Test Incorporation

If the area and delay estimates meet the design constraints, testability incorporation is explored. The test methodologies selected for implementation are given in the input specification and for simplicity can be

$AREA_t = X_t * Y_t$

where:

$$X_t, Y_t = \begin{cases} Max\ Xt_0, Xt_1 & \sum_{i=0}^{1} Yt_i, \\ \sum_{i=0}^{1} Xt_i, & Max\ Yt_0, Yt_1 \end{cases}$$

Xt_i = X dimension of subtree t_i. Yt_i = Y dimension of subtree t_i.

t(t_0 ,_,t_1) = a tree, t(), with subtrees, t_0 and t_1.

DELAY = Max_i { delay(Max_j{path($reg_j \rightarrow fu_i$)})+ delay(fu_i) + delay(Max_j{path($fu_i \rightarrow reg_j$)}) }.

where:

path(reg, fu) = delay(interconnect to fu) + delay(fanout at fu) + delay(mux at fu).

path(fu, reg) = delay(fanout from fu) + delay(interconnect to reg) + delay(mux at reg).

scan path and BIST. These two methodologies were chosen due to their popularity and ease of implementation. The test incorporation problem is defined as given a design structure and test methodology, implement the test such that the design meets the area, delay and test constraints. The objective is to minimize the test overheads (in area and delay) and maximize the testability of the design solution. Test methodologies can be scan path, BIST or a combination of both methods in a non-uniform approach, as illustrated in Figure 12.8 in the context of our synthesized designs.

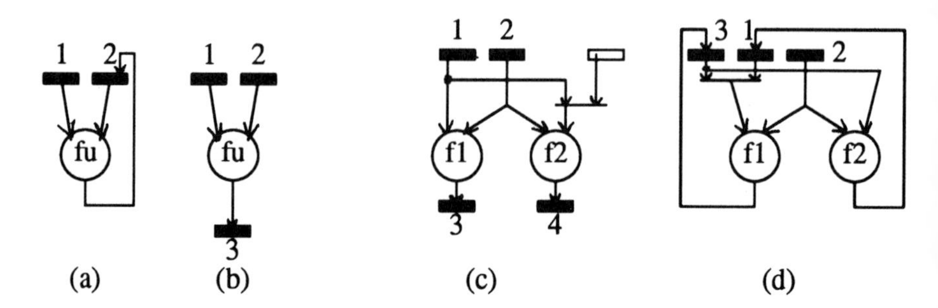

Figure 12.8. a) Scan Path (fu:1,2>2), b) BIST(fu:1,2>3), c) shared BIST (f1:1,2>3; f2:2,1>4), and d) 2-phased BIST (f1:1,2>3; f2:2,3>1) implementation examples for a synthesized data path. Where notation is: (functional unit: left input test register, right input test register > output test register).

The test implementation is done by identifying the circuit to test (ie selecting subtree if nonuniform test methodologies), and then using heuristic rules for assigning or allocating test registers such that all functional units are testable. Test registers could be serial shift registers for scan or LFSR for BIST. The heuristic rules aim to minimize the overhead in additional multiplexor extensions, additional multiplexors, or additional test registers. These three cases are shown in figure 12.9. This would consequently decrease the delay and area overheads (Tsui, 1986) required to meet the design specifications. It is currently assumed for simplicity that the test methodologies are implemented to allow test patterns to be applied to, generated at, or observed at, the inputs or outputs of all functional units, similar to other approaches (Abadir, 1985) except their combinational units are called kernels. In other words all functional units have indirect (through shift registers and multiplexors) controllable and observable inputs and outputs. Design partitioning, naturally represented in the binary tree data structure, is used to implement multiple scan chains, test scheduling, and nonuniform test incorporation of these methodologies.

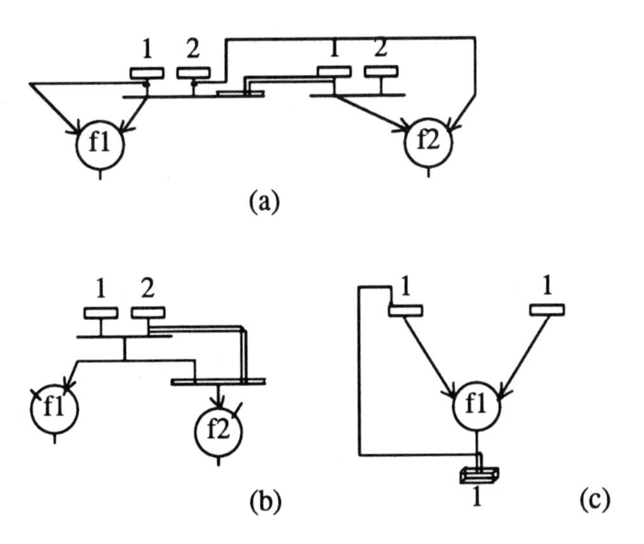

Figure 12.9. Examples of a) multiplexor extension overhead, b) multiplexor overhead, and c) test register allocation (shown as double lines) required during test incorporation stage of CATREE. Both (a) and (b) illustrate input test register assignment for a scan path test implementation. In (c) output test register allocation is illustrated for a BIST test implementation. All cases cause test overheads affecting area and delay.

Each test methodology will provide a different set of rules for assigning test registers. Due to conflicts between register use among functional units, the allocation of new multiplexor extensions, multiplexors, and test registers (which become new leaves in the tree) can be done, as shown in figure 12.9. Single output and input registers of functional units are ranked highest in being transformed into test registers for that functional unit. Next test registers are assigned from small to large sized multiplexors at inputs or outputs of functional units. Finally additional test registers are allocated if necessary. Using these 3 rules, multiplexor extensions are allocated before multiplexors or test registers, thus providing lower overheads (Tsui, 1986) .

The implementation of multiple scan chains for a scan path or nonuniform test methodologies uses the same algorithm outlined above except it is applied to subtrees representing different partitions of the design. For example a double scan chain methodology applies the algorithm to the functional units located in each of the two subtrees located one level below the tree root. If insufficient registers exist in a subtree, then the registers located outside of the subtree are assigned as test registers. When all design registers have been assigned, new test registers could then be allocated. Similarly for nonuniform test incorporation, different test methodologies can be applied to the two subtrees. For example scan path could be implemented in one subtree and BIST could be implemented in the other functional unit subtree. The registers and functional units located in a subtree are highly connected and therefore will be transformed into test registers with small overheads.

The definition of single or multiple scan chains is obtained after test register assignment. The scan chain definition refers to the one bit wide interconnect between the scan registers. It is required to shift patterns on and off the chip for scan path and BIST methodologies. The objective of scan-chain definition is to minimize the interconnect length between registers. This in turn will provide smaller scan delay overheads as demonstrated in (Agrawal, 1984) , where scan definition after layout provided better performance than definition during logic design. A bottom up traversal of the tree or subtree structure is done listing test registers as they are encountered. Scan chains are thus formed from the lower left to the upper right corners of the floorplan, due to the 2-D definition chosen in the tree. This order provides good results for the two examples used in this paper, however other orderings could be defined using the two-dimensional information of the data structure. The tree is ideally suited for these computations due to its natural partitioning and two-dimensional characteristics.

Test scheduling can also be implemented using the tree data structure. Test scheduling refers to the schedule for testing groups of functional units sequentially in more than one test phase for BIST. An example is shown in figure 12.8d), for two phased BIST. This algorithm works bottom up collecting subtrees of X or less functional units, where X represents the number of phases required to fully test the design. Each of the X functional units will be tested during separate phases. Reasonably low overheads are obtained since the functional units found within a subtree are highly connected to their local registers and therefore suited to sharing them.

The partitioning and two-dimensionality of the binary tree data structure provides global information which aids in test incorporation, test scheduling, and scan chain definition.

12.4.5 Feedback

New area, speed, and test cost estimates are obtained after test incorporation. The area and delay estimates use the same algorithms outlined earlier on the current binary tree with test incorporation. A multiple criterion performance measure (Abadir, 1985) is used to determine how close the design solution is to meeting the desired constraints. In this way the 'best' design is updated and stored during the design exploration.

For illustration purposes the testability constraint estimation includes estimation of the fault coverage and test time of the VLSI design. Furthermore it is assumed that each functional unit is combinational logic and is characterized in the VLSI data base with a measure of the total number of faults, and for each test technique, the fault coverage and the number of test vectors or pseudo-random sequences. We use this approach to illustrate the methodology similar to (Abadir, 1985) .

In order to estimate the fault coverage we make the following assumptions. First, faults in the scan registers are all detectable (ie using alternating 1's and 0's test vectors (Agrawal, 1984)). We can also obtain the number of faults detected for each functional unit, recorded for a set of test vectors in our data base (or calculated previously using an automatic test pattern generator (Agrawal, 1984)). The remaining undetected faults result from faults in the (non-test) registers not in the scan chain and multiplexors. For simplicity we assume that the faults traversed in the multiplexor during test mode are detectable. This provides a higher score for multiplexors exercised in more than one mode similar to (Abadir, 1985) where this circuitry is described as being tested for free. Furthermore since faults cannot be propagated properly through multiplexor control logic (Agrawal, 1984) this seems to be a reasonable assumption. One can then estimate the fault coverage globally by summing the total faults detected and dividing by the sum of the total faults in the design, as shown in figure 12.10. The test time is the sum of the time for the scan register testing followed by the time for applying the test vectors or sequences to all functional units. The following equations illustrate the test estimations.

FAULT COVERAGE = Fd/Tf.

where: $\mathrm{Fd} = \{\sum_{i=0}^{m} \mathrm{Fd}(mux_i) + \sum_{i=0}^{f} \mathrm{Fd}(fu_i) + \sum_{i=0}^{r} \mathrm{Fd}(reg_i)\}.$

$$\mathrm{Tf} = \{\sum_{i=0}^{m} \mathrm{Tf}(mux_i) + \sum_{i=0}^{f} \mathrm{Tf}(\,fu_i\,) + \sum_{i=0}^{r} \mathrm{Tf}(\,regs_i)\}.$$

Fd(x) = # of faults detected (by a test method) in the hardware unit x.

Tf(x) = # of total faults present in the hardware unit x.

TEST TIME = $Max_c\ \{\ L_c * Max_i(\ Tv_{(i,c)}\ )\ \}$:for scan chain

TEST TIME = $\sum_{x=0}^{P} Max_i\ (\ Tv_{(i,x)}\ ) + L_0$. :for BIST.

where: L_c: length of (or number of bits in) scan chain c (c>=0).

$Tv_{(x,y)}$: number of test vectors (or sequences) for fu_x on chain y (or active in phase y).

In cases where the number of design registers before test is greater than the minimum number of test registers required, a different number of assigned test registers will vary the fault coverage and test time estimates. Also if multiple scan chains or different test phases are implemented the test time will vary. Finally depending upon how the test methodology was implemented the additional multiplexor extension, multiplexor, or test registers required will also vary the estimation of fault coverage in all cases and the test time in the last case.

In CATREE different test methodologies are applied in an attempt to satisfy the area, delay, and test cost constraints. If the test cost constraint has been met, the strategies outlined for feedback after stage two are used. This creates feedback path F3. However if the test cost constraint

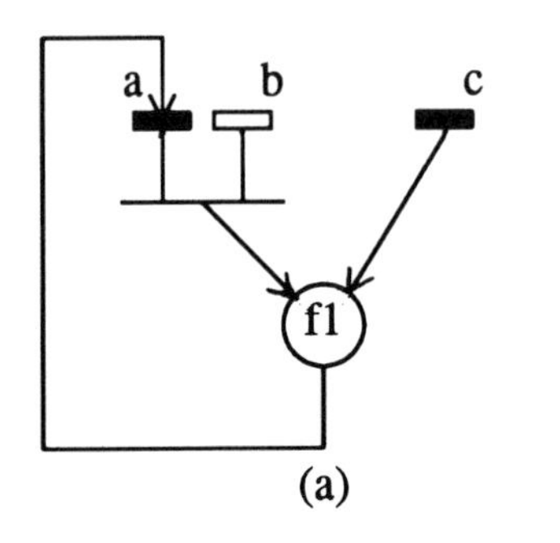

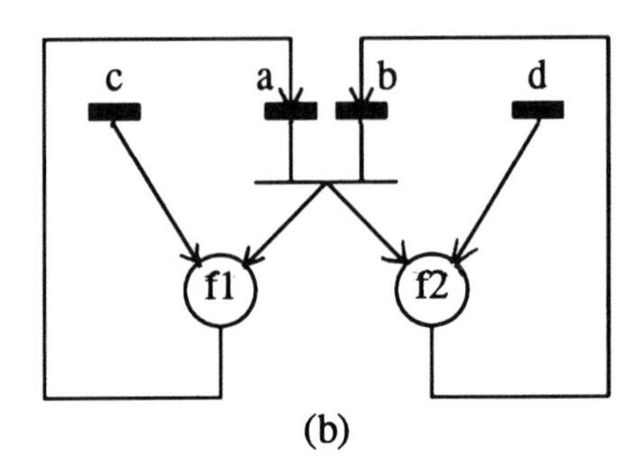

tr = a,c.

Fd= Fd(f1)+Tf(tr)+0.5*Tf(mux).

Tf= Tf(f1)+Tf(tr)+Tf(b)+Tf(mux).

Tt= Tv(f1)*(S(tr)).

(c)

tr = a,b,c in ϕ1 for f1
tr = a,b,d in ϕ2 for f2.

Fd= Tf(tr)+Tf(mux)
+Fd(f1)+Fd(f2).

Tf= Tf(tr)
+Tf(mux)+Tf(f1)+Tf(f2).

Tt= 3*(S(tr))+Tl(f1)+Tl(f2)

(d)

Figure 12.10. An example of fault coverage (Fd/Tf) and test time (Tt) estimation in (c) and (d), for synthesized designs in (a) a scan path and in (b) a double-phased BIST implementation respectively. Where: $f(tr) = \sum_{tr} f(tr)$, S(reg)=# of bits in register; Tl(fu)=# of clock cycles to test fu with BIST; Tv(fu)=# test vectors to test fu with scan path; tr = the test registers.

is not met another test methodology or variation is reimplemented using feedback path F2. When all test methodologies are exhausted and the constraints are still not met feedback path F3 is used to resynthesize the design similar to F2 discussed in section 12.4.3. This provides wide design exploration for testable designs.

12.5 CATREE SYNTHESIS RESULTS

Two examples presented for synthesis in (Paulin, 1987, Pangrle, 1987) were used to illustrate the VLSI synthesis with testability constraints. The first example performs a differential equation using 11 operations. The second example is the elliptical wave filter previously introduced and synthesized by OASIC (see part III, chapter 10). In figure 12.1, feedback path F1 representing the design synthesis search with area and delay constraints has been analyzed in several papers (Thomas, 1983, McFarland, 1986) . We will concentrate on showing results for feedback paths F2 and F3 of figure 12.1. Feedback path F2 illustrates the test methodology search of one design driven by area, delay and test cost whereas feedback path F3 shows the exploration of testable design solutions driven by the area, delay and test cost constraints.

To illustrate the effects of these feedback paths on the testable synthesized design search, four criteria: fault coverage, test time, area, and delay values; were estimated, normalized, weighed, and summed (Breuer, 1985, Zhu, 1988) to obtain design scores for each solution. This design score with equal weights is the multiple criteria performance value also used to select and store the 'best' design found in the search in case all constraints cannot be met. The assumptions made to calculate these four parameters are given in (Gebotys, 1989) . We used the definition of score in (Breuer, 1985) , except our requirement vector was the poorest value of each attribute from all the design solutions. Values were normalized between 0 and 100, and the sum of all weights chosen was one. For example a weight of 4411 represents assigning weights .4, .4, .1 and .1 to fault coverage, test time, area and delay respectively. Thus the highest score represents the best design solution for those weights.

Figure 12.11 shows the results of applying five different test methodologies to one VLSI design solution of the differential equation. This corresponds to feedback path F2 of figure 12.1. Weights 0011 show that the scan path test methodology is the best when strictly area and delay are considered. Equally weighed parameters show that the scan path with two chains approach performs the best. However if the test cost is weighed most important as in score 4411, the shared BIST method becomes the best solution. An example of the CATREE floorplan and register-transfer level solution is found in section 12.8, figure 12.11c,d), where it is compared with CATREE2.

Figure 12.12 and 12.13 show the CATREE differential equation design search with five test methodologies implemented for each solution. Actual testable design details for D0 through D4 are in (Gebotys, 1989) . Figure 12.12 shows the equally weighted area and delay based design scores. Both the design before test incorporation and the testable design are graphed. In figure 12.12, the best solution for the design search based upon area and delay only before test incorporation is design D4, shown by the solid line. When test cost, area, and delay are equally weighted, in figure 12.13, design D3 with the single chain scan path is the best solution. Design D3 is also the best solution for all other test methodologies except the double chain scan path.

Figures 12.14 through 12.16 show the CATREE synthesized design solutions with test incorporation for the EWF. The multipliers were assumed to be two cycle pipelined multipliers. Testable design solution details for the EWF can be found in (Gebotys, 1987) .

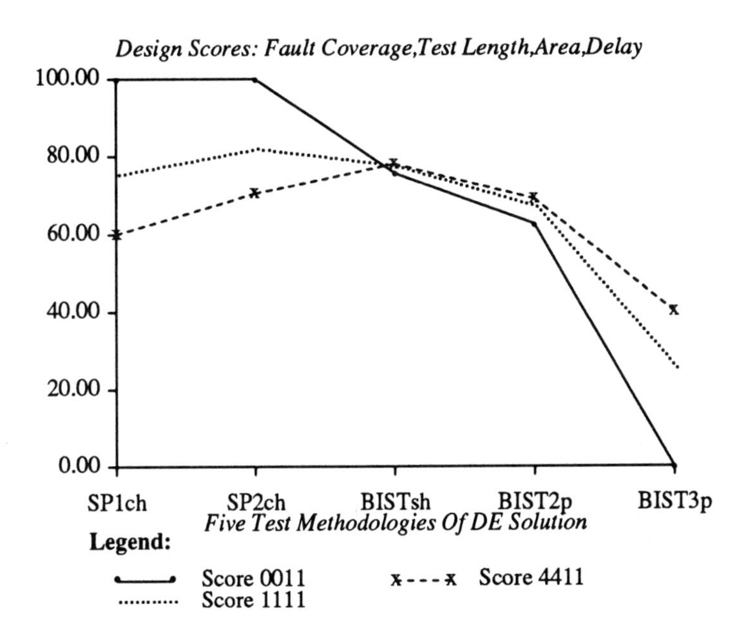

Figure 12.11. Weighted scores of one design solution of differential equation (DE) with five different test methodologies (SP1ch: scan path with 1 chain, SP2ch: scan path with 2 scan chains, BISTsh: BIST shared, BIST2p: BIST with 2 test passes, BIST3p: BIST with 3 test passes).

Figure 12.14 shows the design scores for five test methodologies applied to one EWF design solution. The scan path with two chains is the best testable design solution over all different weights. The second best solutions vary according to which element of the four measures is most important. The single chain scan path is desirable when area and delay are most important while the two phased BIST solution is preferable when test cost is more important.

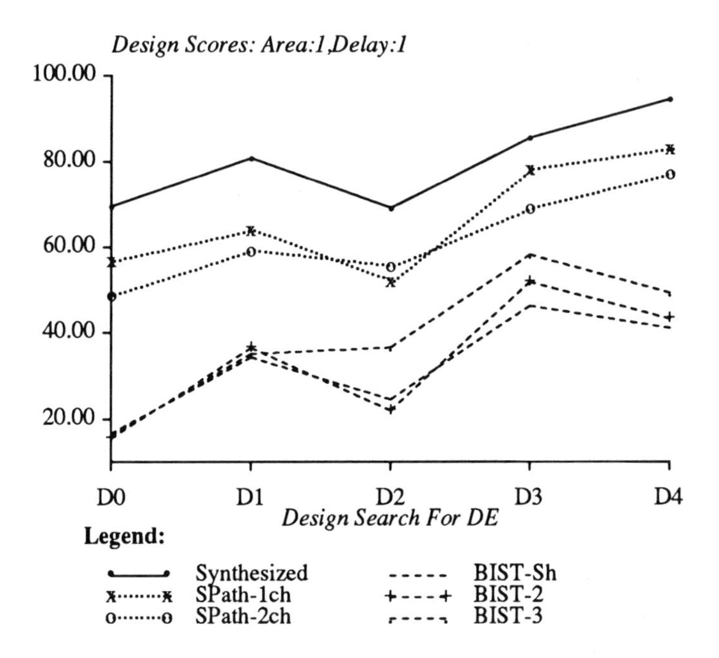

Figure 12.12. Design Scores for equally weighted area and delay only of differential equation example. Synthesized refers to the solution before test incorporation.

Figure 12.15 and 12.16 shows design scores for the four EWF design solutions based upon equally weighted area and delay, and equally weighted area, delay, fault coverage, and test time. Both solutions before test and after test are graphed. f4 is the best solution when only area and delay are considered before test incorporation, shown by the solid line of figure 12.15. In figure 12.16 the best test methodology over all filter design solutions is the double chain scan path. The second best is the 2 phased BIST methodology. For all test methodologies in figure 12.16, solution f4 is the best testable design solution except in the case of the 3

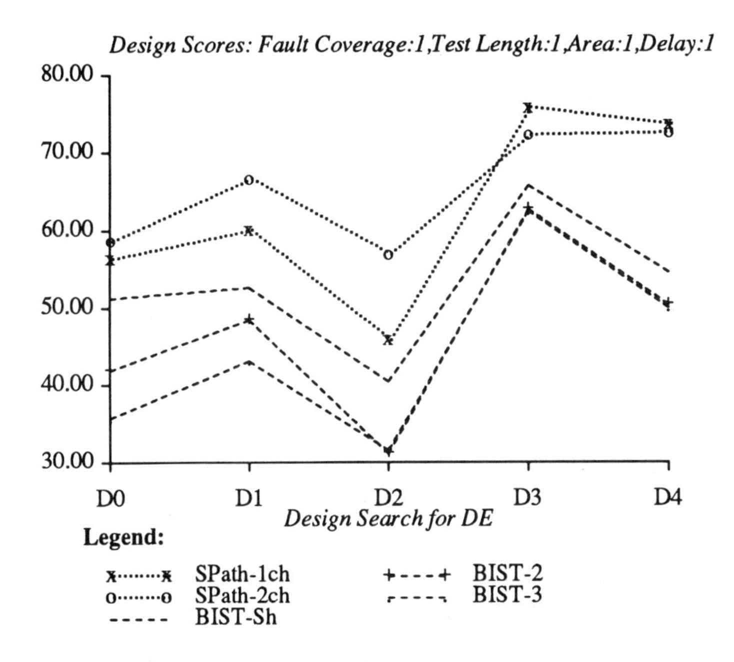

Figure 12.13. Design Scores for equally weighted fault coverage, test length, area, and delay measures are shown for the five testable design solutions of the DE search.

phased BIST implementation. The best testable design solution for this later test methodology is f3. This is due to a higher delay and lower test score associated with the 3 phased BIST test implementation of design f4 as compared to f3.

12.6 CATREE DISCUSSION

Several observations can be made from the differential equation and elliptical wave filter examples. When equally weighted parameters or only area and delay parameters are analyzed the scan path methodology appears to be the test methodology to explore. Another observation is

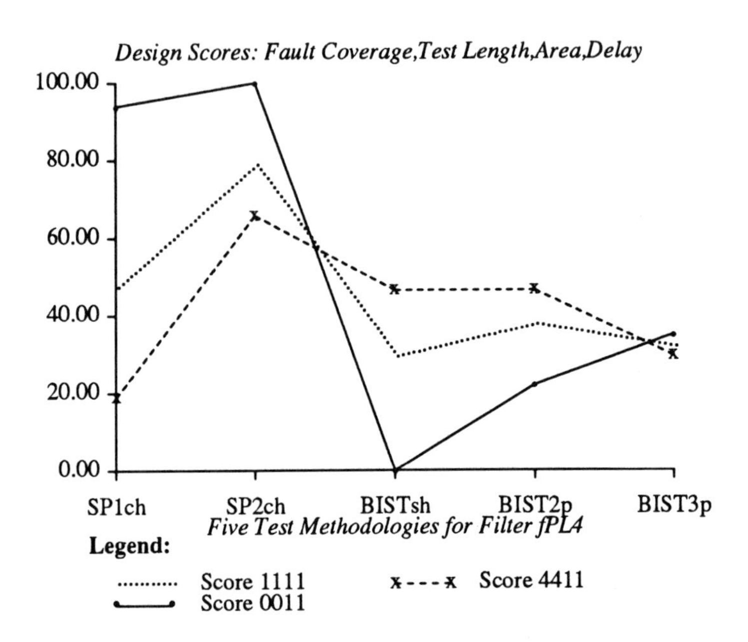

Figure 12.14: Weighted scores of one filter design solution, fPL4, with five different test methodologies.

that weighing the test cost higher produces 'best' testable design solutions using the BIST methodologies. In both cases the implementation details, such as number of chains or test schedule, will vary in each example, hence test implementation of a particular methodology should be fully explored. Also results will vary depending upon the specific library (or library file) the synthesizer is targeted to. Before test incorporation, a synthesis search, based upon area and delay only, produces D4 and f4 (as shown in figure 12.15) as the 'best' solutions of each example respectively. However when test methodologies are implemented and all four parameters are weighed equally, the 'best' testable design solution is different in the differential equation example. This is

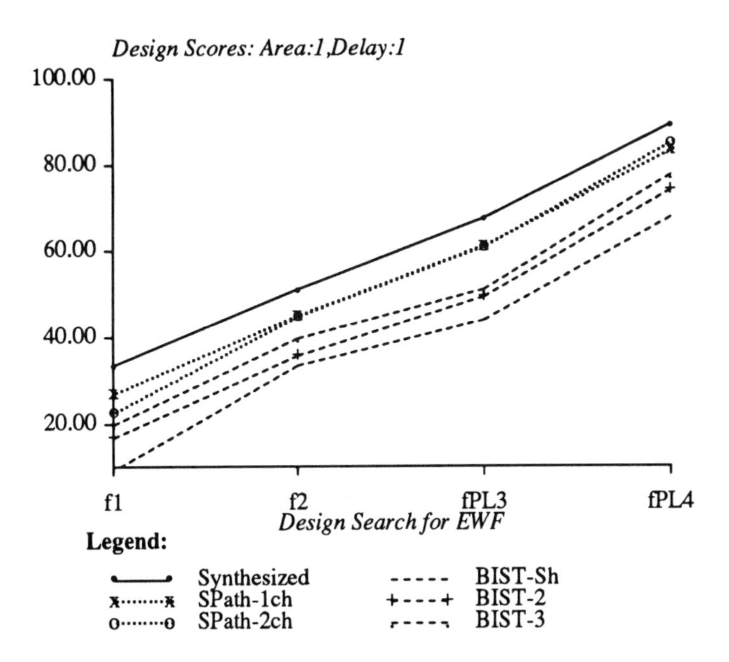

Figure 12.15: Design Scores of equally weighted area and delay only for the elliptical wave filter (EWF) example.

attributable to area and delay, as well as fault coverage and test time, in all test methodologies except the double chain scan path. In the elliptical wave filter example the 'best' testable design solution is different, f3, for one BIST test methodology and the same, f4, in all other test methodologies (as in figure 12.16). For this example the difference in solutions is attributable to fault coverage and delay.

In general we have seen that the best solution for the area-delay based synthesis alone does not always lead to the best testable design solution for all test methodologies. Thus feedback after test incorporation to the synthesis process, feedback path F3, is important for finding

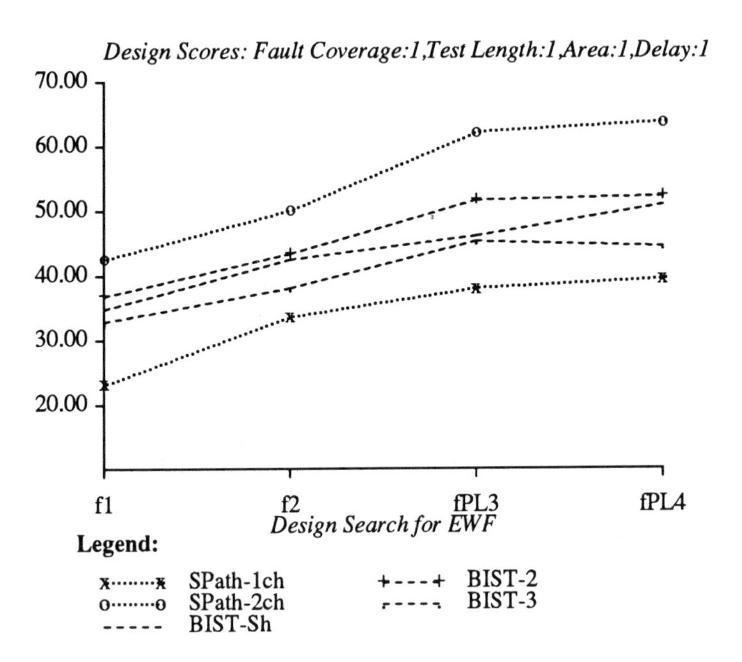

Figure 12.16. Design Scores for equally weighted fault coverage, test length, area, and delay measures for the filter synthesized design search.

good testable design solutions. Secondly, the best synthesized design solution from which the testable design is obtained will often vary for different test methodologies. Hence all test methodologies must be considered, feedback path F2, for all possible design synthesized solutions in a search for a testable design satisfying area, delay and test cost constraints.

The actual weights assigned to the four parameters will directly effect which design becomes the best solution. In practise the weight values must be assigned according to which parameters are the most critical, which in turn would depend upon the application.

If a test-characterized data base library were not available then the test cost parameters, fault coverage and test time, would have to be determined from test software such as an automatic test pattern generator. Test cost could also be extended to include test confidence, input/output pin counts or other attributes (Zhu, 1988) . If the test cost constraint cannot be met after exhaustive test implementations then one could use or create a more testable functional unit or interface to test software for fault coverage improvement or for decreasing the test time. Note as in section 12.4.4, the assumption concerning implementation of the test methodology, may cause overheads to be significant such that all constraints cannot be satisfied. In these cases interface to test software which for example selects a minimum number of nodes for the scan path or which includes only hard to test functional units on scan chains would be necessary. Could also extend CATREE to schedule test phases of non uniform test methodologies instead of restricting this to scheduling for BIST alone.

The floorplanning algorithm produces reasonable results for the design examples used, however other more sophisticated algorithms may improve results for other design examples. The use of a partitioning algorithm (Kernighan, 1970) during tree formation and floorplanning (Gebotys, 1988c) , may improve the test costs (for example by reducing test time), for larger design examples as done in CATREE2 discussed next.

12.7 CATREE2 DESIGN SYNTHESIS STAGES

The stages for CATREE2 are shown in figure 12.2. The main difference between this methodology and CATREE methodology is that the test incorporation is performed after functional unit allocation and binding. In CATREE, test incorporation is performed after a full solution is synthesized. The first stage of CATREE2 involves forming the tree data structure. OASIC again performs the scheduling and allocation of functional units. The test incorporation stage involves functional unit partitioning, test operation assignment and test scheduling. The finish design allocation block in figure 12.2 involves register and interconnect allocation followed by floorplanning and scan chain definition. These allocation algorithms, unlike CATREE, use both design and test information. The area, delay, and test estimation stage is the same as in CATREE. A new feedback procedure using weighted rule sets was also used in CATREE2.

12.7.1 Tree Formation and Functional Unit Binding

The input specification is similar to figure 12.3 except the test methodology specified as scanx or bistx, where x is the number of chains or test phases. Shared BIST and nonuniform test methodologies, discussed for CATREE, were not considered for CATREE2, however could easily be incorporated. A better parser was developed in CATREE2 to handle multiple operation expressions, which are reduced into a series of single operation expressions, and if-then-else code segments. This conditional code is separated into one main trace (including the 'if' true code) and other traces (composed of the 'else' code). Currently arcs are placed between the normal code and conditional code segments for the scheduler to prevent code motion requiring bookkeeping. The scheduler and functional unit allocation is OASIC.

A binary tree data structure is formed from all operations. The min-cut partitioning algorithm (Kernighan, 1970) is recursively applied to form a balanced operation tree based upon a computed score between each pair of operations. The score is obtained from a sum of assigned weights for the same operation type, shared variables, and non-conflicting firetimes.

score(Operation1, Operation2) = $\sum_{i=1}^{4} w_i * r_i$

where:

r_i = # of times rule r is valid.

w_i = weight of rule r.

Operation1 and Operation 2 are the nodes of the graph and the score is the weight on the edge between nodes in the partitioning algorithm. Weights are defined in table 12.3.

Table 12.3. Rule set for Operation Partitioning.

Rule #	Rule Description	Weight
1	shared variables	2
2	same operation	5
3	different times in same trace	5
4	exists in different traces	1

This approach finds trees with the OASIC minimum number of functional units faster than basing the tree formation on variable connectivity only as in CATREE. Functional unit binding uses a bottom up tree traversal algorithm to assign functional unit nodes to subtrees containing non-conflicting operations whose functionality exists in the library. Operations from different traces can be clustered together thus providing functional units for mutually exclusive operations.

12.7.2 Test Incorporation

Test incorporation occurs after the functional unit allocation in CATREE2. This stage involves the three following three tasks: 1) repartition the functional unit tree, 2) add test operation leaves, and 3) perform test scheduling. Test is incorporated into the binary tree such that the next stage which completes the design synthesis, uses both design and test information to make synthesis decisions.

The partitioning algorithm (Kernighan, 1970) is recursively applied to form a balanced mincut tree of functional units based upon their variable connectivity. This provides a balancing of functional units useful for test scheduling, multiple chain definition (even number of chains), or nonuniform test incorporation. The mincut formation is also useful for the tree allocation algorithms which attempt to produce uniformly distributed registers and a minimum number of long interconnect.

A test operation leaf is added to each functional unit subtree, as shown in figure 12.17. The test operations have test input and output variables which together describe the test mode behavior of the design. The test variable names are numerically assigned however their lifetimes are obtained from the test methodology and test schedule (for BIST). The test operation leaf holds the test operation and its input and output test variables. It is assumed for simplicity that the test methodologies are implemented to allow test patterns to be applied to, generated at or observed at, the functional unit inputs or outputs.

For scan path test methodologies all test operations are assigned the same firetime. The input test variable lifetimes do not overlap with the output test variable lifetime. This allows an output test variable to share the same register as an input test variable as is commonly done in scan path testing (McCluskey, 1986) . For multiple scan chains, the subtrees located below the root node of the tree are used to partition the design into groups of functional units which will be tested on separate scan chains.

Test scheduling is required for an x phased BIST implementation. Test operation firetimes are assigned by a bottom up tree traversal algorithm which collects subtrees of x or less functional units. Each test

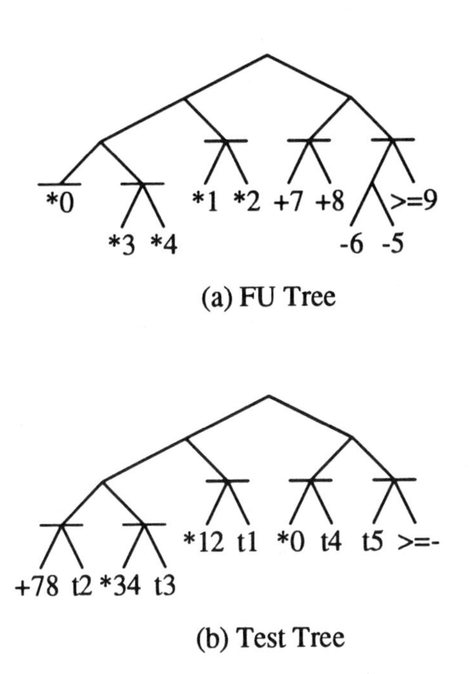

Figure 12.17. An example of CATREE2's functional unit tree before test incorporation in a) and the test tree, after test incorporation in b). Five horizontal lines represent 5 functional unit subtrees (*0,*34,*12,+78,-56>=9). The test operations are tx, where x=1,...,5.

operation of functional units in the subtrees are assigned a different test firetime, corresponding to a different test phase in which the functional unit will be tested. The BIST test input and output variable lifetimes are defined to overlap. This ensures the input pseudo-random number generator or LFSR is a different 'test register' than the output signature analyzer during the same test phase (Williams, 1983).

12.7.3 (Test) Register and Bus Binding

Stage 3 involves the register and interconnect allocation and floor-planning. Both design and test information are used to make synthesis decisions during allocation. In addition the allocation algorithms differ from CATREE by using a voting strategy to determine the best possible clusterings at nodes of the tree. This voting is based on scores between the possible candidate clusters, which are calculated from a weighted rule set.

Register allocation uses a bottom up tree traversal algorithm with conflict and weighted cluster rules applied to test variables and operation output variables of leaf nodes. Each cluster rule has a weight assigned to it, shown in table 12.4, that is used to compute a vote to determine which cluster a variable (or cluster) will be placed (or merged). Cluster rules were used to allow further exploration of the design space. By changing the weights assigned to each rule one can change the synthesized allocations and also observe the effect the rules have on the synthesized solutions. This is a more flexible and interesting method than the CATREE fixed and ordered heuristic allocation rules, however it is a longer process and computationally more expensive. Conflict rules prevent variables from clustering, for example, when two variables have conflicting lifetimes in the same trace. Variable lifetimes in all traces must be checked for conflicts. Variables with lifetimes in different traces will not conflict and can be clustered together. A final clustering with input only variables is done at the tree root at the end of the algorithm.

Each cluster of variables representing a register is then placed into the tree closest to the functional units to which it is most connected. This attempts to maintain the mincut characteristic of the tree. The algorithm tends to produce registers which are evenly distributed among functional units and used in both the normal and the test mode of operation.

Table 12.4. Register allocation cluster rules.

Rule #	Rule Description	Weight
1	same side inputs to a fu	5
2	test variable with inputs for same fu	10
3	output variables of a fu	2
4	test variable with output for same fu	10
5	test variable cluster with test variable	10
6	test variable cluster with normal variable	0
7	same value constants	10
8	same size variables	10

Interconnect allocation involves a top down tree traversal algorithm using conflict and weighted cluster rules on cross variables (xvars) located at different nodes in the tree. Cross variables represent transfers of a variable to a register from a functional unit or from a register to a functional unit. They are stored at nodes in the tree as defined in CATREE. Cross variables of different traces will not conflict with one another. Currently a random or a uni/bi-directional bus topology can be allocated. The later two bus styles are specified in a configuration file which is then used to select appropriate conflict rules. An example of the cluster rules is shown in table 12.5.

Table 12.5. Interconnect cluster rules for CATREE2 synthesis.

Rule #	Rule Description	Weight
1	xvars share a register or fu	30
2	xvars used in different traces	5
3	xvars is regs $\rightarrow$ same fu on same side	5
4	xvars is same reg $\rightarrow$ fus	5

Multiplexor allocation can be done in a local, distributed, or 2 level configuration. The aim is to minimize the number of multiplexor inputs. The local configuration is the same algorithm as described for CATREE in section 6.4.2. The distributed configuration attempts to share multiplexors of registers between two functional units only if this decreases the number of multiplexor inputs is. The algorithm searches for multiplexors of the same type of registers. The 2 level configuration attempts to share an extra level of multiplexors whose output is input to two other multiplexors feeding functional units. Both algorithms are heuristic and attempt to decrease the total number of multiplexor inputs.

The tree algorithm allocates interconnect paths for use in both test and normal modes of operation and is aimed at producing a minimum number of long interconnect.

12.7.4 Feedback

The floorplanning algorithm (Gebotys, 1987) currently assigns alternating x and y dimensions at each level in the tree. Low and high subtrees of each node are also assigned based upon their connectivity to their neighbor nodes. This transforms the binary tree into a two dimensional binary tree as in CATREE.

Area, delay and test cost are computed in stage 4 and resynthesis is invoked if constraints are not met. These three parameters are computed as in CATREE (Gebotys, 1989) from the two dimensional tree and a library containing area, delay and test information for functional units, multiplexors and registers.

Feedback can be performed automatically by using a bottom up tree traversal algorithm that attempts to merge functional units through rescheduling. It also may be done by changing the set of weights assigned to operation, variable or cross variable cluster rules, as shown in figure 12.18. This later method can be used to perform design tradeoffs between test and area. For example if a high weight is assigned to the cluster rule allowing test variables to cluster with other test variables then a minimum number of test registers would most likely be allocated. In figure 12.18b) there are two test registers because t2 clustered with t1. This would decrease the overall area but also decrease the fault coverage if there were additional registers not used as test registers. By changing this weight to a lower value, the number of test registers would increase and consequently the fault coverage would increase. In figure 12.18a) there are three test registers since t2 clustered with c. Hence these weights can be used to explore design tradeoffs. Feedback can also be done manually through rescheduling or adding arcs in the input specification.

12.8 CATREE2 EXPERIMENTS

A differential equation example previously presented for research (Paulin, 1987, Pangrle, 1987) was used to illustrate the CATREE2 design and test synthesis and to compare with results from the CATREE methodology. Figure 12.19 show the CATREE2 and CATREE synthesized design solutions (the OASIC scheduler was not used). A random topology, local multiplexor configuration and double scan chain (scan2) test methodology were implemented in both cases. The design solutions had 5 functional units and 10 test registers. CATREE2 solution had 13 multiplexors or 30 multiplexor inputs. CATREE produced a solution with higher area (it has 31 multiplexor inputs) and a poorer delay than the CATREE2 solution. The critical path had an extra multiplexor input delay and an extra fanout delay. This extra interconnect was required due to the register allocation objective in design synthesis. Minimization of the number of design registers caused an output variable to be allocated to register y1 thus creating an extra multiplexor input. However during test incorporation in CATREE, more test registers than the minimum number of design registers were required, causing the previous

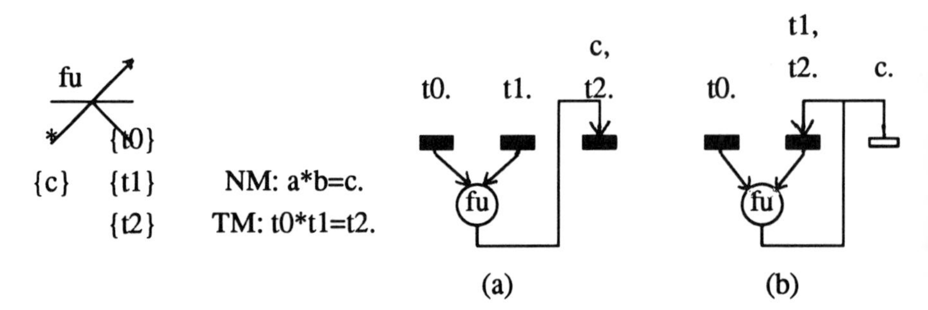

Figure 12.18. An example of the resultant different synthesized designs made by changing rule sets for register allocation. The normal mode (NM) and test mode (TM) behavior and partial tree are shown. In a) the weight of the rule for combining test variables with other test variables is zero, to maximize testability. In b) this rule is given a high weight to decrease the number of test registers, or to minimize the area (c is not a test register).

objective to be unnecessary. Thus the overhead was created since design synthesis was performed without test consideration. Since CATREE2 considered both design and test simultaneously a better solution was obtained.

Figures 12.20 and 12.21 illustrate the decisions CATREE2 can make since both design and test information are used simultaneously during synthesis. In other test incorporation methods (Abadir, 1985, Craig, 1988) which deal with structural information only, these decisions can not be made and therefore in the examples shown larger overheads are produced when they are unnecessary, as illustrated by CATREE2s solutions.

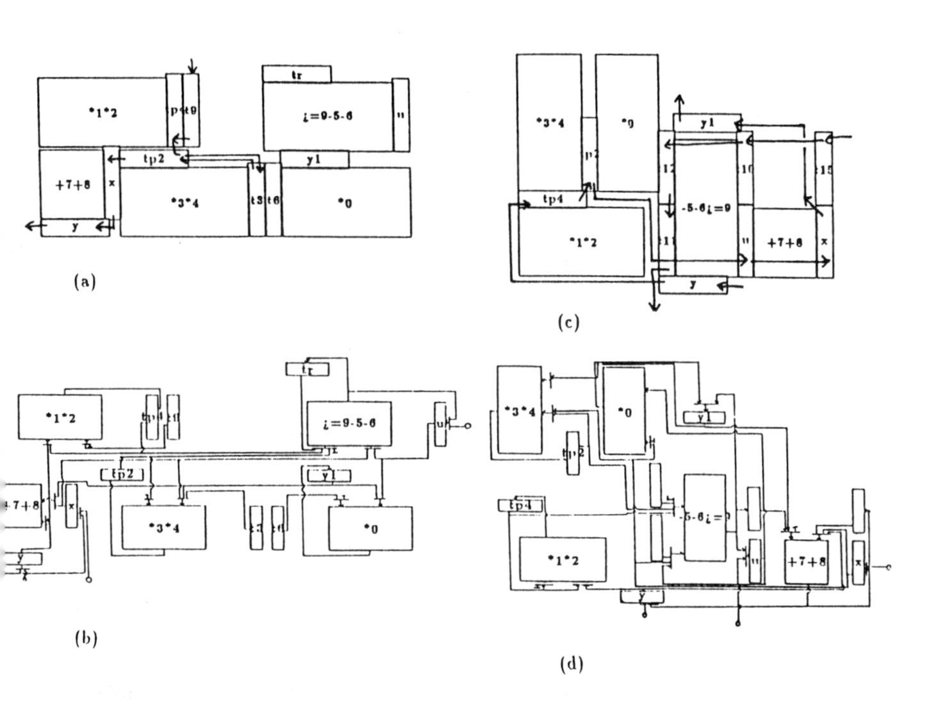

Figure 12.19. An example of a double scan chained synthesized design solution from CATREE2 in a,b) and CATREE in c,d).

Table 12.6 compares the CATREE2 runtimes (rtime), and the number of registers (reg) and busses (bus) of different test methodologies (scanx: x chains, bistx: x phases) with CATREE2 synthesis with no test incorporation (none). All solutions had 5 functional units and implemented a bus topology with weighted cluster rules for a minimum number of test registers. The average runtime overhead for test incorporation was 36%. The runtime values of table 9 are in cpu seconds for CATREE2 written in Quintus Prolog running on a Sun 3/260.

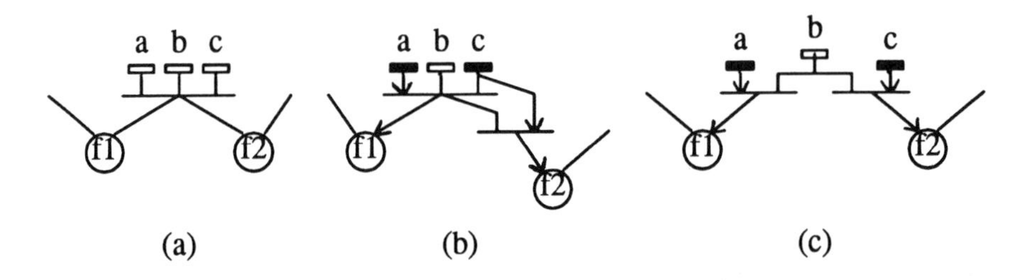

Figure 12.20. Problem = to test both functional units (f1,f2) in parallel: a) obtained from synthesis before test methods, cannot perform a parallel test (without hardware changes) because the 3 register multiplexor is shared; b) the hardware changes, required for parallel testing of a), produce area and delay overheads; and c) the better solution obtained by CATREE2, (preventing the sharing of the multiplexor during synthesis due to test conflicts).

Table 12.7 shows the runtimes (rtime) and the number of operations (K, from the input algorithm) of various CATREE2 synthesized examples. All solutions used a bus topology and BIST test methodology implementation. The average runtime per operation is approximately 8 cpu seconds.

12.9 CATREE2 DISCUSSION

In summary, CATREE2 provided a better solution to the differential equation example than the CATREE methodology because test was considered earlier during the design allocation. This simultaneous design and test methodology provided a higher degree of test and normal mode sharing of registers and interconnect. In this example CATREE2 provided lower area and delay values than CATREE. CATREE required test register allocation in addition to test register assignment. However another advantage to CATREE2 is its ability to perform better design decisions by considering test simultaneously with design, as was shown

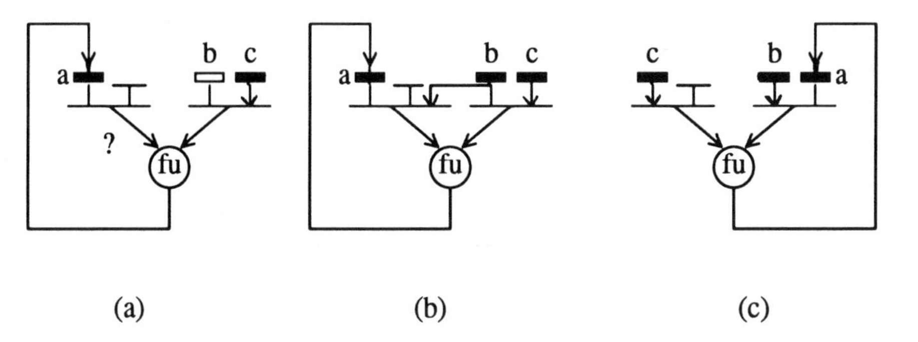

Figure 12.21. Problem = to test the (commutative) functional unit (fu) using a BIST method (where test registers are shown as shaded boxes): a) obtained from synthesis before test incorporation, has no left input test register, see "?"; b) produces area and delay overhead to incorporate test in a) after synthesis; c) is the better solution obtained from CATREE2 which synthesizes left and right test registers for the functional unit.

Table 12.6. CATREE2 testable design solutions for a diffeqn example.

test:	none	scan1	scan2	bist1	bist2	bist3
rtime	48	70	75	80	80	80
regs	6	10	10	15	9	7
bus	8	11	9	9	12	10

in figure 12.20 and 12.21. Without this test knowledge design improvements done during design synthesis may be eliminated during test incorporation or may possibly incur larger performance penalities on the final design (without resynthesis).

CATREE2 runtimes are quite reasonable and do not place a significant burden on the synthesis process by simultaneously considering test. If we use the runtime per operation values to extrapolate, it can be

Table 12.7. CATREE2 runtime versus number of operations(K).

Designs:	test	trace1	diffe	filter
K	5	8	10	34
rtime	35	53	77	345
rtime/K	7	7	8	10

expected that for a larger realistic algorithm, for example, one thousand operations, CATREE2 would perform design synthesis in approximately a few cpu hours. This seems to be a reasonable amount of time making further design explorations feasible. Furthermore fully programmable code does not pose a problem for CATREE2 synthesis since the runtime would approximately be proportional to the total number of operations, with the additional time required in the parser only for extracting traces.

The test incorporation for full controllability and observability of all inputs and outputs of functional units may cause test overheads to be significant such that design constraints cannot be met. A modification so that only the hard-to-test functional units are used for the test mode (thus requiring fewer test operations) could be made in CATREE2. Otherwise interface to test software or use of more testable functional units may be necessary as discussed in CATREE section 6.6.

Assuming that each functional unit has the same number of test vectors, CATREE2 creates a minimum test time by partitioning the tree into equal amounts of functional units. However often the number of test vectors for different functional units will vary as studied in (Craig, 1988) . Thus an extension to this research would be to minimize the test time for functional units with different test times. For example a smaller chain of functional units would be formed if these units required long test times. We could modify CATREE2 to form the final functional unit tree by partitioning based upon a weighted sum of attributes as in the operation tree formation. The attributes would now be connectivity and test lengths. For example two functional units with similar test lengths, and therefore suitable for the same scan chain, would have a higher attraction value. Where as units with different test lengths, belonging on separate chains would have small attraction values.

Also CATREE2 has assumed that each fu has associated with it a set of test vectors (for ScanPath) or a pseudo-random sequence length (for BIST) and a fault coverage. These values were assumed to be precharacterized by test tools such as (Wagner, 1987) . However this could be extended in the following way. Often different test vector sets or different polynomials (or seeds even for pseudorandom testing (Wagner, 1987) could exist for one functional unit in scan path or BIST applications. Thus the CATREE2 problem would be extended to choose the proper set of vectors or best polynomial to achieve the design and test constraints. This extension towards a more realistic design and test library is illustrated in table 11. One final extension would then be to introduce different implementations of functional units as illustrated in the table 12.8 by k. Thus partitioning functional units, assigning test registers and choosing test sets or polynomials would become more complicated and interesting to investigate. However a larger design and test space would be explored.

In this chapter we have described two methodologies, CATREE and CATREE2, which integrate test incorporation with architectural synthesis. Both approaches illustrated the impact of test on area and delay curves of the VLSI architectural design exploration phase. In the next chapter we will make some concluding remarks to summarize the material presented in this text and discuss some future extensions for VLSI testable architectural synthesis.

Table 12.8. Extension of Pre-characterized CATREE(2) Libraries

Library	FU	Library Attributes
CATREE(2)	fu(i)	tv(i),fc(i),a(i),d(i)
extension	fu(i,[k])	tv(i,j,[k]),fc(i,j,[k]), a(i,[k]),d(i,[k])
fu()	═══════	functional unit
tv()	═══════	test vectors (ScanPath) or # of test sequences for a fixed polynomial (BIST)
fc()	═══════	(estimated) fault coverage
a()	═══════	area
d()	═══════	propagation delay
j	═══════	different test set (scanpath) or different polynomial(bist),
fu(i,[k])	═══════	k versions of fu(i)

PART V: SUMMARY AND FUTURE RESEARCH

13.

SUMMARY AND FUTURE RESEARCH

Separate summaries of the OASIC and CATREE methodologies, and concluding remarks (extensions of this research and future CAD tools) are presented in this final chapter of the text.

13.1 OASIC

The OASIC synthesized architectures, which have demonstrated all of the constraints presented in chapters 6 through 9, have been presented in chapter 10. The first three examples (differential equation, elliptical and kalman filter) were taken from the high level synthesis benchmark database. These examples along with the neural network (taken from literature outside of architectural synthesis field) and time constrained examples (also synthesized in chapter 10), provided a good basis for demonstrating the efficiency and flexibility of the OASIC synthesizer. The results of chapter 10 demonstrate how well OASIC can synthesize architectures from input algorithms that range from having a large amount (neural network algorithm) to a small amount (elliptical wave filter) of regularity. And more importantly results indicate that OASIC can efficiently and optimally synthesize architectures even in the presence of complex interface constraints. Other synthesis benchmarks were not selected because many were control dominated, or in other words they were more suited for controller synthesis. For example the architecture itself could be trivially obtained from a few operations in the specification, whereas the remaining operations are control operations useful for controller synthesis.

There exist few published complete architectural solutions to the EWF, a benchmark for over three years; most likely due to its complicated interconnections. For the first time we can directly allocate busses at a very early stage (simultaneously with scheduling) and optimally synthesize an architecture. Furthermore we have shown in chapter 10 that previous heuristic synthesizers (Paulin, 1989) , (Devadas, 1989) , (Lagnese, 1989) have not produced globally optimal schedules and allocations.

Although the worst case complexity is exponential, we have found that many problems (ie Chapter 10) can be solved very fast to global optimums. As algorithms become larger we can take advantage of the mathematical flexibility of OASIC to model systems with hierarchy and regularity (such as the kalman filter and neural network algorithms) using a small number of variables (as demonstrated with functional pipelining). Larger algorithms may be partitioned into smaller code segments at k_{sep} (which maintains optimality) or at other csteps (using partially relaxed OASIC with only objective and other variables εZ). Other partitioning techniques may also be easily incorporated. However more importantly we have demonstrated that over 100 code operations (ie unrolled filter) can be simultaneously scheduled and allocated in very fast cpu times.

This ability to synthesize large complex algorithms is a significant contribution to the synthesis field.

The use of node packing facets provided integer solutions in 16 times faster (defined as CPU time for solving IP without node packing facets divided by the CPU time for IP solution using facets, ie. CPU improved from 600s to 36s) in some cases than the use of previous scheduling constraints (Baker, 1974, Lee, 1989) . Knapsack facets used to improve the lower bounds on the number of busses provided 5 times faster CPU times (30m to 6m). The tightening of constraints was also shown in some cases to improve CPU performance by 1.6 times (420s to 267s).

With respect to previous synthesized architectures, using the cost function (cf) (Devadas, 1989) , we obtained architectures with a 7% to 18 % improvement in area-delay. These improvements come from the optimized number of registers and busses. Furthermore these results are stable over variations in the cost parameters for these two resources. Even when we generously compare with register file architectures we obtain improvements in area-delay.

By scheduling simultaneously with general bus allocation in OASIC we can obtain better interconnect efficient architectures than previous heuristic synthesizers which allocate busses and multiplexors after the schedule is fixed (see figure 10.2).

We can simultaneously solve interface constraints with scheduling and allocation to optimally make tradeoffs between architectural area, speed, and controller size. This is very important since the analysis has a significant effect on the area and delay of the final architecture as demonstrated in this thesis. This is unlike other approaches which schedule after and independently of functional unit allocation (Ku, 1989a) or heuristically readjust the schedule only for functional units (Nestor, 1990) . Secondly the OASIC computation time does not vary significantly in the presence of timing constraints, such as the analog interface (see table

10.21 and 10.8).

We can easily minimize registers in the presence of conditional code unlike synthesizers which use heuristic algorithms (Kurdahi, 1987) . Secondly because our model is mathmatically driven we can minimize the execution times of different mutually exclusive paths of code simultaneously with bus and register allocation, unlike some other synthesizers (Camposano, 1991) which cannot allocate these resources.

List Of Contributions

This section will outline the contributions to the high level synthesis field, high level design methodologies, and systems design methodologies. In addition the impact of this research on extensions for higher level transformations is outlined.

Research Contributions To Architectural Synthesis.

The contributions of this research to the architectural synthesis field are outlined below. The original and new research is highlighted in the following points. For the first time we have:

1. Formulated a model for simultaneous scheduling and allocation of functional units, registers, and busses (Chapters 6 through 9).
2. Solved for globally optimal schedules and allocations which minimize an area and delay piecewise linear cost function (Chapter 10).
3. Formulated timing constraints for interfaces to analog and asynchronous processes (Chapter 9).
4. Applied integer programming facetial techniques to solve architectural synthesis (Chapters 6,7,9,10).
5. Demonstrated that regularity and hierarchy of DAGs can be used to optimally synthesize architectures (Chapter 10).

Contributions To High Level Design Methodologies.

The OASIC methodology has a significant impact on higher level methodologies. Not only does the OASIC model for interfaces support mixed analog/digital systems design, but also its use of regularity and hierarhcy is important. The OASIC mode helps to define how the higher level design automation tool will interact with the user. For example an interactive environment to enable analog/digital partitioning of the behavior, high level code motion, regularity identification, and changes in hierarchy should be part of the future DA tools. The direct impact of changes to the code can be examined, without the nuisance of the heuristic analysis which often do not make use of available fine grain parallelism in the input code. For the first time one can determine whether the "code optimizations" are in fact optimal or not with respect to the parallelism they are supposed to generate.

In summary this research synthesizes in practical execution times, and supports complex constraints including asynchronous/analog interfaces, bus calculation and area-delay cost functions. This is important for industry since its has been estimated it would take an effort larger than the synthesis itself to modify a self synchronized synthesized design to interface with other hybrid processes in a systems design. Also interconnect is seen as the key to high performance architectures and early decisions made during synthesis have the greatest impact on the final design. For the first time we can obtain exact globally optimal schedules and allocations for area-delay cost functions very early in the design cycle. Previous synthesizers could at best guarantee a locally optimal solution (which may not meet area-delay constraints) and could not handle asynchronous/analog interfaces (simultaneously with scheduling and allocation). Finally we have demonstrated (see Chapter 10) that OASIC can handle input algorithms with different types of structure, with over 100 input code operations, and with complex constraints. In summary this research provides globally optimal synthesized architectures, synthesizes large input algorithms in practical execution times, and supports complex constraints and cost functions.

13.2 CATREE

The preliminary research tools, CATREE and CATREE2, are aimed at providing a framework for integration of synthesis and testability. Previously research in these two areas have largely remained separate. These two new VLSI integrated design and test synthesis methodologies with area, delay, and test cost constraints, provide wider design exploration, early performance feedback including test cost, and integration of design and test considerations. Preliminary results show that good testable design solutions are produced by simultaneously synthesizing the design's behavioral and test modes. This earlier consideration of test seems to make better design choices during allocation and allows confident design improvements taking both design and test information into consideration.

This has implications for work in both research and industrial environments. In industry, such a tool, when given an algorithm and test methodology, would provide wider exploration of solutions, with area, delay and test cost constraints satisfied earlier in the VLSI design cycle. Also valuable design cycle time would be saved by providing wider exploration of design solutions with area, delay and test cost constraints satisfied earlier in the VLSI design cycle. In research, it has been shown that testability should be considered an important part of the design synthesis search along with area and delay. Thus feedback path F3 of CATREE, along with F1 and F2, should be used when synthesizing designs for an algorithmic description. Also the tight coupling of the design and test synthesis, provided by CATREE2, can be used as a basis for further exploring improvements in design searches with area, delay, and test cost constraints.

In summary the CATREE and CATREE2 design synthesizers provide a useful approach towards incorporating testability within the synthesis process using a two dimensional binary tree data structure. The tree data structure provides allocation algorithms of reasonable complexity without sacrificing quality (Gebotys, 1987) . Integration of the VLSI data base is achieved through a common binary tree data structure used in complete synthesis with testability unlike a combination of separate synthesis and testability tools such as (Granacki, 1985, McFarland, 1986) tools with (Abadir, 1985, Zhu, 1988) . The binary tree data structure also supports automated feedback to the synthesis process after test exploration. Finally the two dimensionality naturally represented within the binary tree data structure supports testability incorporation and area-delay estimation, unlike other structures lacking this property (Abadir, 1985, Zhu, 1988) . CATREE has been implemented in Quintus Prolog and is available as part of the Waterloo VLSI CAD Tool package (Elmasry, (null)) .

13.3 FUTURE EXTENSIONS

The following is a list of future extensions for OASIC and CATREE. These are divided into extensions for the model and extensions for the solution strategy.

Model Extensions

An extension of the OASIC model for register transfer architecture such as SPAID (Haroun, 1989) could be performed. In this case one would minimize the size of register files, number of busses, and numbe r of multiplexors.

It would be useful to extend the bus allocation constraints for $t \geq 3$, since our current OASIC model (chapter 7) only provides an exact number of busses for t=1,2.

An extension to allocate and select functional units representing chained operations and subsequently change the clock period would be very useful. Currently we can select chained operations, however we cannot account for different clock periods.

OASIC could also be extended to allocate storage and busses for different data sizes. Another extension for OASIC would be solution of the simultaneous scheduling, allocation and binding problem, where the number of bus drivers is modelled.

Test incorporation performed simultaneously with a second binding optimization phase similar to (Gebotys, 1990) would be another interesting extension. Future research would also include extending the synthesis to fully programmable algorithmic input using trace scheduling techniques to optimize hardware implementation for high probability traces. Also completing the front end parser to extract traces, implementing the trace scheduling approach, extending automatic feedback strategies, including controller costs, and automating testability measures earlier in the synthesis process, ie. evaluating test measures for chained operations, would be investigated in the future.

Solution Strategies Extensions

It would also be interesting to study the use of branch and bound on partial orders instead of variables of the model. We would expect good results since there are in some applications fewer partial orders than variables and the subpolytope of precedence constraints is very tight.

Investigation and generalization of other facets for a branch and cut automated tool would also be another extension. The development of a heuristic strategy for selection of variables to branch on would also improve CPU times for architectural synthesis. This was discussed briefly in OASIC (chapter 10) where branching on the most constrained variables (the multiplication) provided significant CPU improvement.

The use of the node packing decomposition technique could also be investigated for only a subset of OASIC ($x_{i,j,k}$). Preliminary results found that all variables were set to 0.5. This was similar to other results in (Grimmett, 1985) . Secondly since bounds are known to be very poor with this model (Padberg, 1973) and results with the facet model were very good, this approach was not pursued in this thesis, however could be further analyzed in future research.

13.4 CONCLUDING REMARKS

For the first time we have tightly integrated architectural synthesis with testability, which should have a large impact on decreasing the VLSI design cycle. In addition this research has for the first time formulated a complete IP model for simultaneous scheduling and allocation including an exact allocation of busses. Secondly we are the first to apply facetial techniques to solve architectural synthesis. We have further shown that globally optimal architectures can be synthesized in faster CPU times than previous research. This is very important for industry because a mathematical basis (OASIC) is used for synthesis which supports correct architectures through formal verification of only the model and not the architectural solution as required by other heuristic synthesizers. The OASIC synthesizer uses very reliable and robust mathematical software which has been proven in other applications over a number of years. This model forms the basis of a CAD tool which can be brought to market very quickly because of this tested software. This research is important since it has shown that the use of mathematical theory, developed over the last 25 years, can have significant impact on solving new VLSI problems. Secondly since our solution is mathematically driven we can easily support complex constraints and a wide range of different types of input algorithms, in contrast to previous heuristic approaches. The mathematical basis of the solution strategy combined with engineering creativity will enhance our ability to extend our technique to a wide range of problems in VLSI.

REFERENCES

Abadir, M.S. and M.A. Breuer, ''Constructing Optimal Test Schedules for VLSI Circuits Having Built-In Test Hardware,'' *International Symposium on Fault Tolerant Computing*, pp. 165-170, 1985.

Abadir, M.S. and M.A. Breuer, ''A Knowledge-Based System For Designing Testable VLSI Chips,'' *IEEE Design and Test*, pp. 56-68, 1985.

Agrawal, V.D., ''Information Theory in Digital Testing - A New Approach to Functional Test Pattern Generation,'' *Int'l Conf. on Computers and Circuits*, pp. 928-931, 1980.

Agrawal, V.D., S.K. Jain, and D.M. Singer, ''Automation in Design For Testability,'' *Custom Integrated Circuits Conference*, pp. 159-163, 1984.

Agrawal, V.D., K. Cheng, and P. Agrawal, ''CONTEST: A Concurrent Test Generator for Sequential Circuits,'' *Design Automation Conference*, pp. 84-89, 1988.

Aho, A.V., J.E. Hopcroft, and J.D. Ullman, *The Design And Analysis of Computer Algorithms*, Addison-Wesley, 1974.

Baker, K.R., *Introduction to Sequencing and Scheduling*, John Wiley & Sons, 1974.

Balakrishnan, M., A.K. Majumdar, D.K. Banerji, J.G. Linders, and J.C. Majithia, ''Allocation of Multiport Memories in Data Path Synthesis,'' *IEEE Transactions on Computer-Aided Design of Integrated Circuits and Systems*, vol. 7, no. 4, pp. 536-540, April 1988.

Balakrishnan, M. and P. Marwedel, ''Integrated Scheduling and Binding: A Synthesis Approach For Design Space Exploration,'' *Design Automation Conference*, pp. 68-74, 1989.

Balraj, T.S. and M.J. Foster, ''Miss Manners Silicon Compiler for Synchronizers,'' *Advances in Research in VLSI at MIT*, 1986.

Barbacci, M.R., ''Instruction Set Processor Specification (ISPS): The Notation and Its Application",'' *IEEE Transactions on Computer*, vol. C-30, pp. 24-40, 1981.

Beausang, J. and A. Albicki, "The Design For Testability Process: Definition and Exploration," *Int'l Conf. on Computer Design*, pp. 362-365, 1987.

Bhandari, I., M. Hirsch, and D. Siewiorek, "The Min-Cut Shuffle: Toward a Solution for the Global Effect Problem of Min-Cut Placement," *ACM/IEEE Design Automation Conference*, pp. 681-685, 1988.

Bhatt, S.N., F.R.K. Chung, and A.L. Rosenberg, "Partitioning Circuits for Improved Testability," in *Advanced Research in VLSI, Proceedings of 4th MIT Conference*, pp. 91-106, MIT Press, 1986.

Bondy, J.A. and U.S.R. Murty, *Graph Theory with Applications*, North Holland, 1976.

Borriello, G. and R.H. Katz, "Synthesis and Optimzation of Interface Transducer Logic," *Int'l Conf on Computed Aided Design*, pp. 274-277, 1987.

Borriello, G. and E. Detjens, "High-Level Synthesis: Current Status and Future Directions," *Design Automation Conference*, pp. 477-482, 1988.

Breuer, M.A. and X. Zhu, "A Knowledge based System for selecting a test methodology for a PLA," *Design Automation Conference*, pp. 259-265, 1985.

Brewer, F. and D. Gajski, "Chippe: A System for Constraint Driven Behavioral Synthesis," *IEEE Transactions on Computer Aided Design*, vol. 9, no. 7, 1990.

Brooke, A., D. Kendricke, and A. Meeraus, *GAMS/MINOS Users Manual*, Scientific Press, 1988.

Brucker, P., M.R. Garey, and D.S. Johnson, "Scheduling Equal-Length Tasks Under Treelike Precedence Constraints to Minimize Maximum Lateness," *Mathematics of Operations Research*, vol. 2, no. 3, August 1977.

Brzozowski, J.A. and C-J.H. Seger, "Advances in Asynchronous Circuit Theory - Part I: Gate & Unbounded Inertial Delay Models," *Bulletin of the European Association for Theoretical Computer Science*, p. pgs 52, Oct 1990.

Camposano, R., "Path Based Scheduling for Synthesis," *IEEE Transactions on Computer Aided Design*, pp. 85-93, 1991.

Carley, L.R., D. Garrod, R. Harjani, E. Ochotta, and R.A. Ruttenbar, "ACACIA The CMU Analog Design System," *Research Report No. CMUCAD-89-64*, November 1989.

Chandra, S.J. and J.H. Patel, "Experimental Evaluation of Testability Measures for Test Generation," *IEEE Transactions on Computer-Aided Design*, vol. 8, no. 1, pp. 93-97, 1989.

Chen, T. and M.A. Breuer, "Automatic Design for Testability Via Testability Measures," *IEEE Transactions on Computer-Aided Design*, vol. CAD-5, no. 1, pp. 3-11, 1985.

Chen, X. and M.L. Bushnell, "A Module Area Estimator for VLSI Layout," *IEEE Design Automation Conference*, pp. 54-59, 1988.

Clocksin, W.F. and C.S. Mellish, *Programming In Prolog*, Springer-Verlag, 1984.

Cloutier, R.J. and D.E. Thomas, "The Combination of a Scheduling, Allocation, and Mapping in a Single Algorithm," *Design Automation Conference*, pp. 71-76, 1990.

Coffman, E.G., *Computer and Job Shop Scheduling Theory*, J.Wiley and Sons, 1976.

Cook, B., Solving General Integer Programming, Talk at University of Waterloo, Canada, May 1990.

Craig, G.L., C.R. Kime, and K.K. Saluja, "Test Scheduling and Control for VLSI Built-In Self-Test," *IEEE Transactions on Computer*, vol. 37, no. 9, 1988.

Crowder, H., E.L. Johnson, and M. Padberg, "Solving Large Scale Zero-One Linear Programming Problems," *Operations Research*, vol. 31, no. 5, pp. 803-834, September 1983.

Depuydt, F., G. Goossens, J. Van Meerbergen, F. Catthoor, and H. DeMan, "Scheduling of Large Scale Signal Flow Graphs based on Metric Graph Clustering," *IFIP Conference on High Level Architectural Synthesis and Logic Synthesis*, 1990.

Devadas, S. and A.R. Newton, "Algorithms for Hardware Allocation in Data Path Synthesis," *IEEE Transactions on Computer Aided Design of Circuits and Systems*, July 1989.

Dussault, J.A., "A Testability Measure," *Proc Semiconductor Test Conf*, pp. 113-116, 1978.

Dutt, N.D. and D.D. Gajski, "Designer Controlled Behavioral Synthesis," *Design Automation Conference*, pp. 754-757, 1990.

Ellis, J.R., *Bulldog: A Compiler for VLIW Architectures*, MIT Press, 1986.

Elmasry, M.I., *For Information write to Professor M.I.Elmasry, Director of VLSI Research Group, University of Waterloo, Waterloo, Ontario, Canada, N2L 3G1.*.

Fey, C.F. and D.E. Paraskevopoulos, "A Model of Design Schedules for Application Specific IC's," *Custom Integrated Circuits Conference*, pp. 490-496, 1986.

Fogg, D.C., "Assisting Design Given Multiple Performance Criteria," VLSI Memo No.88-479, Massachusets Institute of Technology, 1988.

Foulds, L.R., *Optimization Techniques: An Introduction*, Springer-Verlag, 1981.

Freeman, S., "Test Generation for Data-Path Logic: The F-Path Method," *IEEE Journal of Solid-State Circuits*, pp. 421-427, 1988.

Fujiwara, H. and S. Toida, "The Complexity of Fault Detection Problem for Combinational Logic Circuits," *IEEE Trans. on Computers*, vol. C-31, no. 6, pp. 555-560, 1982.

Funatsu, S., N. Wakatsuki, and T. Arima, "Test Generation Systems in Japan," *DA Symposium*, pp. 114-122, 1975.

Fung, H.S. and J.Y.O. Fong, "An Information Flow Approach to Functional Testability Measures," *Int'l Conf on Circuits and Computers*, pp. 460-463, 1982.

Fung, H.S. and S. Hirschhorn, "An Automatic DFT System for the Silc Silicon Compiler," *IEEE Design and Test*, pp. 45-47, 1986.

Gajski, D.D. and R.H. Kuhn, "Guest Editors Introduction: New VLSI Tools," *IEEE Computer Magazine*, vol. 16, no. 12, pp. 11-14, 1983.

Gajski, D.D., N.D. Dutt, and B.M. Pangrle, "Silicon Compilation (Tutorial)," *Custom Integrated Circuits Conf*, pp. 102-109, 1986.

Galiay, J., Y. Crouzet, and M. Verigniault, "Physical vs Logical Fault Models," *IEEE Trans on Computers*, vol. C-29, no. 6, 1980.

Garey, and Johnson, *Computers and Intractability,* Freeman and Co., 1979.

Gebotys, C.H. and M.I. Elmasry, "VLSI Design Synthesis Exploration With Testability Constraints," *Tech. Rep. , UW/ICR 87-14, Department of Electrical Engineering, University of Waterloo, Waterloo, Ontario, Canada*, 1987.

Gebotys, C.H. and M.I. Elmasry, "VLSI Design Synthesis with Testability," *Design Automation Conference*, pp. 16-21, 1988a.

Gebotys, C.H. and M.I. Elmasry, "Integrated Design and Test Synthesis," *International Conference on Computer Design*, pp. 398-401, 1988b.

Gebotys, C.H. and M.I. Elmasry, "Integration of Algorithmic VLSI Synthesis with Testability Incorporation," *IEEE Custom Integrated Circuits Conference*, 1988c.

Gebotys, C.H. and M.I. Elmasry, "Integration of Algorithmic VLSI Synthesis with Testability Incorporation," *IEEE Journal of Solid-State Circuits*, vol. 24, no. 2, pp. 409-416, 1989.

Gebotys, C.H. and M.I. Elmasry, "A Global Optimization Approach to Architectural Synthesis," *IEEE International Conference on Computer Aided Design*, 1990.

Gebotys, C.H. and M.I. Elmasry, "A Global Optimization Approach for Architectural Synthesis," UW/ICR 91-01, p. 28pgs, 1991a.

Gebotys, C.H. and M.I. Elmasry, "Simultaneous Scheduling and Allocation for Cost Constrained Optimal Architectural Synthesis," *ACM/IEEE Design Automation Conference*, 1991b.

Gebotys, C.H. and M.I. Elmasry, "High Performance Optimal Architectural Synthesis," *IEEE Custom Integrated Circuits Conf*, 1991c.

Gebotys, C.H., "A Global Optimization Approach to Architectural Synthesis fo VLSI Digital Synchronous Systems With Analog and Asynchronous Interfaces," *Dept of ECE,Univ of Waterloo, PhD Thesis*, July 1991x.

Gill, P.E., W. Murray, and M.H. Wright, *Practical Optimization,* Academic Press, 1981.

Girczyc, E.F., R.J.A. Buhr, and J.P. Knight, "Applicability of a Subset of Ada as an Algorithmic Hardware Description Language for Graph-Based Hardware Compilation," *IEEE Transactions on Computer-Aided Design*, vol. CAD-4, no. 2, pp. 134-142, 1985.

Goel, P., "Test Generation Costs Analysis and Projections," *Design Automation Conference*, pp. 77-84, 1980.

Goldstein, L.H. and E.L. Thigpen, "SCOAP: Sandia Controllability/Observability Analysis Program," *Design Automation Conference*, pp. 190-196, 1980.

Golumbic, M.C., *Algorithmic Graph Theory and Perfect Graphs,* Academic Press, 1980.

Gottlieb, E.S. and M.R. Rao, "The Generalized Assignment Problem: Valid Inequalities and Facets," *Mathematical Programming*, vol. 46, pp. 31-52, 1990.

Granacki, J., D. Knapp, and A. Parker, "The ADAM Advanced Design Automation System: Overview, Planner and Natural Language Interface," *Design Automation Conference*, pp. 727-730, 1985.

Grimmett, and Pulleyblank, "Random Near Regular Graphs and the Node Packing Problem," *Operations Research Letters*, vol. 4, no. 4, 1985.

Grossman,, "Mixed Integer NonLinear Programming Techniques for the Synthesis of Engineering Systems," EDRC-6-83-90, Engineering Design Research Center, Carnegie Mellon University, 1990.

Grotschel, M., "On the Symmetric Travelling Salesman Problem: Solution of a 120 City Problem," *Mathematical Programming Study*, vol. 12, pp. 61-77, 1980.

Gupta, R. and G. DeMicheli, "Partitioning of Functional Modules of Synchronous Digital Systems," *Int'l Conf on Computed Aided Design*, 1990.

Hafer, L. and A. Parker, "A formal Method for the Specification, Analysis and Design of Register-Transfer-Level Digital Logic," *IEEE Transactions on Computer Aided Design of Circuits and Systems*, vol. CAD-2, no. 1, pp. 4-17, Jan 1983.

Hal, L. and D. Shmoys, "Near Optimal Sequencing with Precedence Constraints," *Proc. of Integer Programming and Combinatorial Optimization Conf.*, 1990.

Hammer, P.L., E.L. Johnson, and B.H. Korte, *Annals of Discrete Mathematics 4, Discrete Optimization I,*, North Holland, 1979.

Harjani, R., R.A. Rutenbar, and L.R. Carley, "OASYS: A Framework for Analog Circuit Synthesis," *Res. Rept. # CMUCAD-89-65*, p. 31pgs., Nov 1989.

Haroun, B. and M. Elmasry, "Architectural Synthesis for DSP Silicon Compilers," *IEEE Transactions on Computer Aided Design of Circuits and Systems*, vol. CAD-8, no. 4, April 1989.

Hashimoto, A. and J. Stevens, "Wire Routing by Optimizing Channel Assignments with Large Apertures," *Proc. 8th Design Automation Workshop*, pp. 155-169, 1971.

Hayati, S. and A. Parker, "Automatic Production of Controller Specifications from Control and Timing Behavioral Descriptions," *Design Automation Conference*, pp. 75-80, 1989.

hlsw, and B. Mayo (Coordinator), *High-Level Synthesis Workshop Clearinghouse, email: hlsw-request@decwrl.dec.com*, 1988.

Ho, C.Y., R.T. Jerdonek, S.E. Noujaim, and D. Schumacher, "A High Performance 1.5 Micron CMOS 24X24 BIT Multiplier," *Custom Integrated Circuits Conf.*, pp. 30-33, 1984.

Holton, W.C. and R.K. Cavin, "A Perspective on CMOS Technology Trends," *Proceedings of the IEEE*, vol. 74, no. 12, 1986.

Holton, W.C. and R.K. Cavin, "A Perspective on CMOS Technology Trends," *Proceedings of the IEEE*, vol. 74, no. 12, 1986.

How, M.M. and B.Y.M. Pan, "Amber: A Knowledge-Based Area Estimation Assistant," *International Conference on Computer Design*, pp. 180-183, Oct. 1986.

Huang, C., Y. Chen, Y. Lin, and Y. Hsu, "Data Path Allocation Based on Bipartite Weighted Matching," *Design Automation Conference*, pp. 499-504, 1990.

Hwang, C-T, J-H. Lee, and Y-C. Hsu, "A Formal Approach to the Scheduling Problem in High-Level Synthesis," *IEEE Transactions on CAD*, vol. 10, no. 4, pp. 464-475, 1991.

Ibarra, O.H. and S.K. Sahni, "Polynomially Complete Fault Detection Problems," *IEEE Transactions on Computers*, vol. C-24, no. 3, pp. 242-249, 1975.

IEEE, Computer, *Special Issue on Artifical Neural Systems*, March 1988.

Jain, R., A. Parker, and N. Park, "Predicting Area-Time Tradeoffs for Pipelined Designs," *IEEE Design Automation Conference*, pp. 35-41, 1987.

Jain, R., M.J. Mlinar, and A. Parker, "Area-Time Model for Synthesis of Non-Pipelined Designs," *Int'l Conf on Computed Aided Design*, 1988.

Jain, R., M.J. Mlinar, and N. Park, *Area-Time Model for Synthesis of Non-Pipelined Designs*, pp. 48-51, 1988.

Johannsen, D.L., S.K. Tsubota, and K. McElvain, "An Intelligent Compiler Subsystem for a Silicon Compiler," *Design Automation Conference*, pp. 443-450, 1987.

Karmarkar, N., "A new polynomial-time algorithm for linear programming," *Combinatorica*, vol. 4, pp. 373-395, 1984.

Karmarkar, N., "An Interior Point Approach to NP-Complete Problems," *Proc. of Integer Programming and Combinatorial Optimization Conf.*, pp. 351-366, May 1990.

Kernighan, B. and S. Lin, "An Efficient Heuristic Procedure for Partitioning Graphs," *Bell Systems Technical Journal*, pp. 291-307, 1970.

Knapp, D.W. and A.C. Parker, "A Data Structure for VLSI Synthesis and Verification," *Digital Integrated Systems Center Report, DISC/83-6, Dept of Electrical Engineering, University of Southern California, Los Angeles, CA 90089-0871*, Oct 1983.

Krasniewski, A. and A. Albicki, "Simulation-Free Estimation of Speed Degradation in NMOS Self-Testing Circuits for CAD Applications," *ACM/IEEE Design Automation Conference*, pp. 808-811, 1985a.

Krasniewski, A. and A. Albicki, "Automatic Design of Exhaustively Self-Testing Chips with BILBO Modules," *International Test Conference*, pp. 362-371, 1985b.

Ku, D.C. and G. DeMicheli, "Relative Scheduling Under Timing Constraints," CSL-TR-89-401, Stanford Technical Report, 1989a.

Ku, D.C. and G. DeMicheli, "Optimal Synthesis of Control Logic From Behavioral Specifications," CSL-TR-89-402, Stanford Technical Report, 1989b.

Kuchcinski, K. and Z. Peng, "Parallelism Extraction from Sequential Programs for VLSI Applications," *Microprocessing and Microprogramming*, pp. 87-92, 1988.

Kung, S.Y., H. Whitehouse, and T. Kalaith, *VLSI and Modern Signal Processing*, Prentice-Hall, 1985.

Kung, S.Y. and J.N. Hwang, "Parallel Architectures for aNN," *International Conference on Neural Networks*, 1988.

Kurdahi, F.J. and A.C. Parker, "Plest: A Program for Area Estimation of VLSI Integrated Circuits," *Design Automation Conference*, pp. 467-473, 1986.

Kurdahi, F.J. and A.C. Parker, "REAL: A Program for Register allocation," *Design Automation Conference*, pp. 210-215, 1987.

Lagnese, E.D., "Architectural Partitioning for Systems Level Design of Integrated Circuits," *CMUCAD-89-27, Carnegie Mellon University, PhD Thesis*, 1989.

Langeler, G., "The Last Decade of Design Automation. and the next.," *Design Automation Conference*, 1989.

Lawler, E.L., *Combinatorial Optimization Networks and Matroids*, Holt-Rinehart-Winston, 1976.

Lawler, E.L., J.K. Lenstra, A.H.G. Rinnooykan, and D.B. Shmoys, *The Travelling Salesman Problem, A Guided Tour of Combinatorial Optimization*, Wiley-Interscience, 1985.

Lee, J., Y. Hsu, and Y. Lin, "A New Integer Linear Programming Formulation for the Scheduling Problem in Data Path Synthesis," *Int'l Conf on Computed Aided Design*, 1989.

Leiserson, C.E., F.M. Rose, and J.B. Saxe, *Optimizing Synchronous Circuitry by Retiming*, pp. 87 - 116, 1970.

Lippmann, R.P., "An Introduction to Computing with Neural Networks," *IEEE ASSP Magazine*, pp. 4-22, April 1987.

Ly, T.A., W.L. Elwood, and E.F. Girczyc, "A Generalized Interconnect Model for Data Path Synthesis," *Design Automation Conference*, 1990.

Mann,, "Technologies for aNN," *Custom Integrated Circuits Conference*, 1988.

Marlett, R., "An Effective Test Generation Systems For Sequential Circuits," *Design Automation Conference*, pp. 250-256, 1986.

Marwedel, P., "A new Synthesis Algorithm for the MIMOLA software system," *Design Automation Conference*, pp. 271-277, 1986.

McCluskey, E.J., "Design for Testability," in *Fault-Tolerant Computing*, ed. D. K. Pradhan, vol. I, Prentice-Hall, 1986.

McCluskey, E.J., "Why We Need Test," *Int'l Symposium on Circuits and Systems*, 1990.

McFarland, M., A. Parker, and R. Camposano, "Tutorial on High-Level Synthesis," *Design Automation Conference*, pp. 330-336, 1988.

McFarland, M., A. Parker, and R. Camposano, "The High Level Synthesis of Digital Systems," *Proceedings of IEEE*, vol. 78, pp. 301-318, 1990.

McFarland, M.C., "Using Bottom-Up Design Techniques in the Synthesis of Digital Hardware from Abstract Behavioral Descriptions," *Design Automation Conference*, pp. 474-480, 1986.

McFarland, M.C., "Reevaluating the Design Spae for Register-Transfer Hardware Synthesis," *International Conference on Computer-Aided Design*, pp. 262-265, 1987.

McQueen, C., "A Data Structure for VLSI Layout," *M.A.Sc. Thesis, Dept of Electrical Engineering, University of Toronto, Toronto, Ontario. Canada,* 1984.

Meng, and Brodersen, "Asynchronous Circuit Synthesis," *IEEE Transactions on CAD*, 1989.

Minty, G.J., "On Maximal Independent Sets of Vertices in a Claw-Free Graph," *Journal of Combinatorial Theory*, vol. B28, pp. 284-304, 1980.

Motohara, A., K. Nishimura, H. Fujiwara, and I. Shirakawa, "A Parallel Scheme for Test-Pattern Generation," *International Conference on Computer-Aided Design*, pp. 156-159, 1986.

Nemhauser, G.L. and L.E. Trotter, "Properties of Vertex Packing and Independence System Polyhedra," *Mathematical Programming*, vol. 6, pp. 48-61, 1974.

Nemhauser, G.L. and L.E. Trotter, "Vertex Packings: Strcutural Properties and Algorithms," *Mathematical Programming*, vol. 8, pp. 232-248, 1975.

Nemhauser, G.L. and L.A. Wolsey, *I Conf.nteger and Combinatorial Optimization*, Wiley Interscience, 1988.

Nestor, J.A. and D.E. Thomas, "Behavioral Synthesis with Interfaces," *Int'l Conf on Computed Aided Design*, 1986.

Nestor, J.A. and G. Krishnamoorthy, "SALSA: A New Approach to Scheduling with Timing Constraints," *Int'l Conf on Computed Aided Design*, pp. 262-265, 1990.

Padberg, M.W., "On the Facial Structure of Set Packing Polyhedra," *Mathematical Programming*, vol. 5, pp. 199-215, 1973.

Padberg, M.W., "Covering, Packing, and Knapsack Problems," in *Annals of Discrete Mathematics*, vol. 4, pp. 265-287, North-Holland, 1979.

Padberg, M.W. and S. Hong, "On the Symmetric Travelling Salesman Problem: A computational Study," *Mathematical Programming Studies*, vol. 12, pp. 61-77, 1980.

Pangrle, B.M. and D.D. Gajski, "Design Tools for Intelligent Silicon Compilation," *IEEE Transactions on Computer-Aided Design*, vol. CAD-6, no. 6, pp. 1098-1112, 1987.

Papadimitriou, C.H. and M. Yannakakis, "Analysis of Parallel Algorithms," *SIAM*, 1990.

Park, N. and A. Parker, "SEHWA: A Program for Synthesis of Pipelines," *Design Automation Conference*, 1986.

Parker, A. and N. Park, "Interface and I/O Protocol Descriptions," in *Advances in CAD for VLSI - Hardware Description Languages*, ed. R.W. Hartenstein, vol. 17, pp. 110-113, North Holland, 1987.

Paulin, P., *discussion with Pierre Paulin*, 1987.

Paulin, P.G. and J.P. Knight, ''Scheduling and Allocation For Behavioral Synthesis of Pipelined ASICs,'' *Canadian Conference on VLSI*, pp. 229-234, 1987.

Paulin, P.G., ''Force Directed Scheduling,'' *IEEE Transactions on CAD*, pp. 661-679, 1989.

Peng, Z., ''A Formal Methodology for Automated Synthesis of VLSI Systems,'' *Linkoping Studies in Science and Technology. Dissertations*, no. 170, 1987.

Petersen, B.R., B.A. White, D.J. Salomon, and M.I. Elmasry, ''SPIL: A Silicon Compiler with Performance Estimation,'' *Int'l Conf on Computed Aided Design*, pp. 500-503, 1986.

Pfahler, P., ''Automated Datapath Synthesis: A Compilation Approach,'' *Processing and Microprogramming*, vol. 21, pp. 577-584, 1987.

Ra, and Grossman, ''Relation Between MILP Modelling and Logical Inferences for Chemical Process Synthesis,'' EDRC-06-87-90, Engineering Design Research Center, Carnegie Mellon University, 1990.

Rajan, J.V., ''Automatic Synthesis of Microprocessors,'' *Res.Rept.#CMUCAD89-2, Carnegie Mellon University, PhD Thesis*, 1989.

Ratiu, I.M., A. Sangiovanni-Vincentelli, and D.O. Pederson, ''VICTOR: A Fast VLSI Testability Analysis Program,'' *IEEE Test Conference*, pp. 397-401, 1982.

Rosales, B.C., ''Test and Synthesis: A Critical Coupling,'' *Semicustom Design Guide, High Performance Systems*, p. 61, 1989.

Roth, J.P., W.G. Bouricius, and P.R. Schneider, ''Programmed Algorithms to Computer Tests to Detect and Distinguish between Failures in Logic Circuits,'' *IEEE Trans Electron. Comput.*, vol. EC-16, no. 5, pp. 567-580, 1967.

Sabo, D.G., D. Johannsen, and R. Yau, ''Genesil Silicon Compilation and Design For Testability,'' *Custom Integrated Circuits Conference*, pp. 416-420, 1986.

Sarkar, V., *Partitioning and Scheduling Parallel Programs for MultiProcessors*, MIT Press, 1989.

Sarma, R.C., M.D. Dooley, N.C. Newman, and G. Hetherington, ''High-Level Synthesis Technology Transfer to Industry,'' *Design Automation Conference*, pp. 549-554, 1990.

Savir, J., ''Syndrome-Testable Design of Combinational Circuits,'' *IEEE Trans on Computers*, vol. C-29, no. 6, 1980.

Schrijver, A., *Theory of Linear and Integer Programming*, Wiley InterScience Series in Discrete Mathematics and Optimization, 1986.

Shen, J.P., W. Maly, and F.J. Ferguson, ''Inductive Fault Analysis of nMOS and CMOS Integrated Circuits,'' *SRC-CMU Center for CAD, Res. Rept. CMUCAD-85-51*, 1985.

Shiva, S.G., ''Automatic Hardware Synthesis,'' *Proceedings of IEEE*, vol. 71, no. 1, Jan 1983.

Springer, D.L. and D.E. Thomas, ''Exploiting the Special Structure of Conflict and Compatibility Graphs in High Level Synthesis,'' *Int'l Conf on Computed Aided Design*, pp. 254-257, 1990.

Stok, L., ''Interconnect Optimization During Data Path Allocation,'' *Workshop on High Level Synthesis*, 1989.

Subrahmanyam, P.A., ''A Framework for System Timing,'' in *VLSI Specification, Verification and Synthesis*, ed. P.A. Subrahmanyam, Kluwer Academic Publishers, Boston, 1988.

Susskind, A.K., ''Testing by Walsh Coefficients,'' *IEEE Trans on Computer*, vol. C-32, no. 2, pp. 198-201.

Susskind, A.K., ''Survey of VLSI Test Strategies,'' *Custom Integrated Circuits Conference*, pp. 276-280, 1984.

Thomas, D.E., C.Y. Hitchcock, T.J. Kowalski, J.V. Rajan, and R.A. Walker, ''Automatic Data Path Synthesis,'' *IEEE Computer*, pp. 59-70, 1983.

Treleavan, P., M. Pacheco, and M. Vellasco, ''VLSI Architectures for Neural Networks,'' *IEEE Micro*, pp. 8-42, 1989.

Trickey, H., "Flamel: A High-Level Hardware Compiler," *IEEE Transactions on Computer Aided Design*, pp. 259-269, 1987.

Trotter, L.E., "A Class of Facet Producing Graphs for Vertex Packing Polyhedron," *Discrete Mathematics*, vol. 12, pp. 373-388, 1975.

Tseng, C. and D.P. Siewiorek, "Automated Synthesis of Data Paths in Digital Systems," *IEEE Transactions on Computer-Aided Design*, pp. 379-395, 1986.

Tsui, F.F., *LSI/VLSI Testability Design*, McGraw-Hill, 1986.

Ueda, K., H. Kitazawa, and I. Harada, "CHAMP:Chip Floor Plan for Hierarchical VLSI Layout Design," *IEEE Transactions on Computer Aided Design*, Jan 1985.

Ullman, J.D., "NP-Complete Scheduling Problems," *Journal of Computer and Systems Science*, vol. 10, pp. 384-393, 1975.

Varma, P. and Y. Tohma, "A Knowledge-Based Test Generator for Standard Cell and Iterative Array Logic Circuits," *IEEE Journal of Solid-State Circuits*, vol. 23, no. 2, pp. 428-435, 1988.

Wagner, K.D., C.K. Chin, and E.J. McCluskey, "Pseudorandom Testing," *IEEE Transactions on Computers*, vol. C-36, no. 3, pp. 332-343, 1987.

Walker, R.A. and D.E. Thomas, "Design Representation and Transformation in The System Architect's Workbench," *Research Report CMUCAD-87-34*, August 1987.

Weber, S., "For VLSI, Multichip Modules may become the Package of Choice," *Electronics*, pp. 106-174, Apr 1989.

Wei, R. and C. Tseng, "Column Compaction and Its Application to the Control Path Synthesis," *International Conference on Computer-Aided Design*, pp. 320-323, 1987.

Williams, T.W. and K.P. Parker, "Design for Testability - A Survey," *Proceedings of the IEEE*, 1983.

Wolsey, L.A., "Further Facet Generating Procedures for Vertex Packing Polytopes," *Mathematical Programming*, vol. 11, pp. 158-163, 1976.

Zahir, R. and W. Fichtner, "Making Use of Timing Constraints for Controller Synthesis," *Workshop on High Level Synthesis*, 1989.

Zhu, X. and M.A. Breuer, "A Knowledge-Based System for Selecting Test Methodologies," *IEEE Design and Test*, pp. 41-59, 1988.

INDEX